HOW TO STUDY PHYSICS

by

JUDAH LANDA

Member New York State Regents Physics Committee

Coordinator of Physics, Medical Science Institute, Midwood H.S. at Brooklyn College

Instructor of Physics, Brooklyn College

JAY - EL PUBLICATIONS

Brooklyn, New York

Cover art: Dispersion of light, by *Lorraine L. Carlsen*

Copyright © 1994

JUDAH LANDA

No part of this book may be reproduced in any form without written permission from the publisher.

ISBN 0-9639716-0-3

Manufactured in the United States of America

DEDICATED TO

a truly devoted science educator

DAVID R. KIEFER

Contents

Introduction. So You're Taking Physics	1
***Chapter 1.* General Suggestions**	
1.1 The Use of Words	4
1.2 The Use of Formulas	7
1.3 Units in Physics	10
1.4 Symbols in Physics	16
1.5 Solving the Problems	18
1.6 Concept of Proportionality	27
***Chapter 2.* Motion**	
2.1 Vocabulary and Formulas	36
2.2 Things to Know	37
2.3 Solving the Problems	39
2.4 On Your Own Problems	50
***Chapter 3.* Motion Graphs**	
3.1 Types of Graphs	54
3.2 Formulas	56
3.3 Things to Know	56
3.4 Solving the Problems	61
3.5 On Your Own Problems	73
***Chapter 4.* Vectors**	
4.1 Vocabulary and Formulas	80
4.2 Things to Know	83

4.3	Solving the Problems	84
4.4	On Your Own Problems	111

Chapter 5. **Laws of Motion**

5.1	Vocabulary	115
5.2	Things to Know	116
5.3	Solving the Problems	120
5.4	On Your Own Problems	137

Chapter 6. **Friction**

6.1	Vocabulary	141
6.2	Things to Know	142
6.3	Solving the Problems	143
6.4	On Your Own Problems	153

Chapter 7. **Two-dimensional Motion**

7.1	Vocabulary and Formulas	157
7.2	Things to Know	158
7.3	Solving the Problems	161
7.4	On Your Own Problems	173

Chapter 8. **Gravity**

8.1	Vocabulary and Formulas	178
8.2	Things to Know	179
8.3	Solving the Problems	182
8.4	On Your Own Problems	190

Chapter 9. **Electricity**

9.1	Vocabulary and Formulas	194
9.2	Things to Know	196
9.3	Solving the Problems	199
9.4	On Your Own Problems	207

Chapter 10. **Magnetism**

10.1	Vocabulary and Formulas	210
10.2	Things to Know	214
10.3	Solving the Problems	217
10.4	On Your Own Problems	226

Chapter 11. Momentum Conservation

11.1	Vocabulary and Formulas	233
11.2	Things to Know	234
11.3	Solving the Problems	235
11.4	On Your Own Problems	244

Chapter 12. Forms of Energy

12.1	Vocabulary and Formulas	248
12.2	Things to Know	250
12.3	Solving the Problems	253
12.4	On Your Own Problems	263

Chapter 13. Conservation of Energy

13.1	Vocabulary and Formulas	268
13.2	Things to Know	269
13.3	Solving the Problems	272
13.4	On Your Own Problems	292

Chapter 14. Waves and Particles

14.1	Vocabulary and Formulas	299
14.2	Things to Know	302
14.3	Solving the Problems	307
14.4	On Your Own Problems	318

Chapter 15. Geometric Optics

15.1	Vocabulary and Formulas	323
15.2	Things to Know	326
15.3	Solving the Problems	330
15.4	On Your Own Problems	345

Appendix A.	Solutions	349
Appendix B.	Physical Constants	372
Appendix C.	Trigonometric Functions	373
Appendix D.	Indices of refraction wavelengths of light specific heats	374

Index 375

INTRODUCTION

SO YOU'RE TAKING PHYSICS

I am frequently asked by prospective physics students, "Is physics really as difficult as everyone says it is?" Some time later, after the student has experienced a few tests or quizzes, the question becomes, "What can I do to pass physics?" Colleagues and parents echo the refrain by asking, "Why do so many of the best and brightest students perform so poorly in their physics classes?" And professional educators are forever coming up with new theories to explain why so few high school and college students succeed in the "hard" sciences, with physics usually occupying the number one slot on the list of courses in that catagory.

The fact of the matter is that the typical first time physics taker, irrespective of whether it is on the high school or college level, finds the subject strange and uniquely challenging. No course of study or academic activity engaged in prior to taking physics adequately prepares the student for the physics experience. The science courses typically studied earlier (biology, chemistry and earth science) consist mostly of memorization of facts and procedures - skills that are not of paramount importance to the mastery of physics. The mathematics courses taken before physics may have, if taught properly, included a measure of emphasis on problem solving skills that are helpful in physics. But this is typically not practiced with sufficient intensity to make much of a difference and represents only one aspect of the challenge posed by physics.

The situation is aggravated by the almost total absence of appropriate guidance as to how the study of physics is to be approached. Physics calls for special study techniques and habits. For the first time in their lives students are confronted by a whole matrix of subtle concepts that demand an unusually high level of focused thought and attention to be grasped fully. And the problems to be solved are frequently based on those subtle concepts - rendering the problems doubly challenging. A student who approaches the study of physics as he or she would chemistry, biology, earth science or algebra is bound to become frustrated and disappointed.

This book is designed to fill the gaps in that missing guidance. The book is not titled "Physics Made Easy" for it doesn't purport to achieve that *nor can it* achieve that (no book can). But by providing the necessary and constructive guidance as to how physics is to be studied, this book goes a long way toward removing a major obstacle on the road to success in physics.

The suggestions offered in this book are organized in the following manner. First, generalized techniques to be employed and habits to be cultivated in the study of any area of physics are discussed. Then these suggestions are applied to various specific topics, in each of which is demonstrated how the suggestions can be implemented. For the convenience of the reader each chosen topic is highlighted by a list of words, definitions and formulas essential to mastery of that topic. This is followed by a list of "Things to Know" to help the reader avoid the most common traps and mistakes. Then problems are solved step by step, with the problems ascending the ladder of complexity. The items in "Things to Know" and the steps in "Solving the Problems" accomplish two important things - they help the reader focus on the essential features of the subtle physics concepts relevant to the topic and they repeatedly demonstrate the utility of the generalized suggestions. By seeing over and over again how much of a difference this approach makes, the reader will develop the habit of thinking and acting in the manner prescribed by the generalized suggestions. Success in physics will then be within reach.

Finally, the student is asked in each section to solve a few problems "On Your Own". Just in case some obstacle still exists to hinder his or her ability in solving any of these problems, the step-by-step solutions to some of them (those with an astrisk) are provided in elaborate detail in the answer key at the end of the book. Students should always make sure they can solve the problems *entirely on their own* before walking into a test.

1

CHAPTER ONE

GENERAL SUGGESTIONS

1.1 THE USE OF WORDS

Words in physics are meant to be used with a degree of precision highly unusual in ordinary everyday conversation. Most of us communicate by linking words together without paying careful attention to the precise meaning of our words. Some of the words we use do not even have well defined meanings. This is unfortunate because it is the source of much confusion, misunderstanding, disagreement and even conflict in the affairs of mankind. As most of us go through life using words in the manner dictated by our habits, we find that we must stop every now and then to deal with the consequences wrought by our words - when it's usually too late. Lawyers and editors, for whom words are bread and butter, have long been aware of how important it is to strive to use words precisely - to say *exactly* what one means and to mean *exactly* what one says. This is why when an important document or contract is about to be signed, we usually retain the services of an attorney. We realize that under these circumstances our words must be chosen

General Suggestions 5

carefully for significant consequences may flow from them, but that we are ill equipped - by force of habit - to do so. An expert in the precise meaning and use of words is called in to help us select words that accurately reflect out thoughts and intentions. And despite these precautions disputes still occur all too frequently regarding the intent of one or another clause in a documented agreement.

In the sciences one simply cannot succeed without confronting and addressing this fundamental issue. The problem is particularly acute in physics for two reasons. First, many of the words employed in physics are also used in ordinary conversation with altogether different meanings and connotations. Second, the physics terms have unusually precise and subtle definitions, the nuances of which are easy to overlook if you do not pay careful attention to them. As an example, consider the word "acceleration" - an important and frequently encountered term in physics. In ordinary conversation, acceleration is synonomous with "change in speed". We might say, for example, "the car accelerated" or "the car experienced an acceleration". Both phrases have the same meaning - the car's speed increased. In physics the word acceleration is used in the sense of "acceleration rate", the change in speed *per time*. While we can see with our eyes that a car's speed has increased (the car accelerated), we cannot see its acceleration rate - a mathematical quantity found by dividing the change in speed by the elapsed time.

If you're not careful you might easily be lulled into equating acceleration (for which the symbol a is used) with "increase in speed" (Δv) instead of making the correct association between acceleration and the quantity "change in speed per time" ($\Delta v / \Delta t$).

Many times have I seen students make this erroneous equation between a and Δv. The two quantities frequently have very different values and are always associated with different units. A car that goes "from zero to sixty (m.p.h.) in 8 seconds" experiences a change in speed (Δv) of 60 mph, but its acceleration rate is 7.5 miles per hr per sec.

The word "force" in physics is often thought of as a "push or pull". It is even so defined in some physics textbooks. In everyday conversation, to push or pull something is to make it move in the direction of the push or pull. The American College Dictionary's definition of "push" is "to move by exerting force" or "thrust in order to move away". After using the word force in this manner for many years it is no wonder that students are surprized to discover that in physics an object can be moving one way at the same time that it is being forced another way, even the opposite way. Force in physics is more precisely defined as "an influence applied to an object in an attempt to change its motion status". The attempt to change the motion status may or may not succeed (depending on the presence of other forces) , but the force is still there. And even when the attempt does succeed, the object may still not be moving in the direction of the applied force (see chapter five). Similarly, the physics term "power" is often interchanged with the everyday concept of "force". And the word "work" is confused with "effort".

There is a long list of such words in physics - words with precise and subtle definitions that do not coincide with how they are used in everyday conversation. What is to be done about this? The first step is to be aware of the problem. Second, whenever a physics term is introduced make sure you become informed of its complete, correct and precise definition - something this book will strive to help you achieve. Third, and most important, repeatedly focus upon and make use of the physics terms and their definitions, particularly in contexts (such as problem solving) where use of precise definitions is in demand.

Another helpful device is to "talk physics" with friends, colleagues and others, in order to provide yourself with many opportunities to use the terms correctly and precisely. It helps, for example, to repeatedly use the phrase "acceleration rate" (as opposed to merely "acceleration") and to say "the force *acts* to the right" (as opposed to "the force *goes* to the right"). With frequent use in appropriate contexts the message is absorbed into one's consciousness and eventually becomes internalized.

So we postulate rule number one for success in physics: **USE WORDS WITH PRECISION!**

1.2 THE USE OF FORMULAS

The student of physics is inundated with a multitude of formulas. There is a formula for virtually everything in physics. It is tempting to see this as providing an opportunity for an easy way to success - reduce physics to a list of formulas. When confronted with a physics problem, the student may reason, all one need do is pull the appropriate formula out of a hat, plug numbers into symbols, and out will come the answer.

Nothing could be further from the truth. Students who view physics as a list of formulas soon discover that it is not an effective approach. It also takes the joy and heart out of the subject. The deeper meaning of the great ideas of physics is lost and the student soon finds himself or herself inside a jungle of symbols and rules so numerous that it becomes quite an unenviable task keeping track of them all.

The challenge posed by the many formulas of physics is compounded by the fact that most of the formulas come with strings attached - conditions and restrictions regarding when and how they may be used. In addition, the symbols that appear in a formula inevitably have subtle meanings that may be overlooked only at one's peril. The letter t in motion related formulas, for example, represents *time*. What could be simpler than that, right? But what time does it represent? Not time on the clock, but the time elapsed between two events. Which two events? The formula does not reveal. You must provide that information based on your understanding of the concepts that form the basis of the formula.

Since formulas are a much used tool in physics, they must be important. What then is to be done? What is the best way to handle them? The answer is provided by the fact that behind every formula lurks a story. Every formula comes from some idea or principle and was derived via a certain process. The important thing to do is to understand and appreciate that story - the basis of the formula. Once fully grasped the story is usually easy to remember and provides guidance as to how, when and where the formula is to be used. The story also helps remind us of the meaning of the symbols contained in the formula.

The formula $F = ma$ seems, at first glance, to be extremely simple and easy to use. It consists of only three symbols and contains no complicated mathematical operations to perform. Yet students frequently misuse the formula by neglecting to equate F, the force, with the *net* (unbalanced) force acting on an object (as opposed to merely any force present). In the formula for "work",

$W = Fd\cos\theta$, on the other hand, the letter F represents any force that acts on an object, even if there are other forces present. These contradictory conditions are a chore to memorize and easy to forget - if they are treated as mere details. They are automatically in our minds if we are cognizant of the stories behind the formulas - the ideas they flow from.

The story behind the formula $F = ma$, in brief, goes as follows: According to the first law of motion, if no unbalanced force acts on an object its motion continues forever unchanged. The second law ($F = ma$) is based on the first law. It comes to tell us what changes in motion to expect when an unbalanced (net) force *is* present. If motion continues unchanged in the absence of a net force, the presence of a net force ought to provide change in motion every second that it acts. The difference between a strong net force and a weak one is then manifested by the amount of change in speed each produces per second of acting. The strength of a net force, F, is therefore proportional to the change in speed *per time*, $\Delta v/\Delta t$, it produces - in other words, to the acceleration rate, a, of the object it acts upon. When this is combined with the mass factor we arrive at $F \propto ma$. Since the unit of force, the Newton (N), is conveniently defined as the amount of force that accelerates 1 kg at the rate of 1 meter per sec per sec (1 m/s^2) the relationship between F, m and a becomes $F = ma$.

Appreciating the flow of ideas from the first law of motion to the second law and the other ingredients that go into the making of the second law brings the whole subject to life. The basic outline of these exciting ideas, when fully understood and appreciated, is easily remembered. Approaching formulas this way is much more effective than concentrating on the memorization of seemingly disconnected, lifeless details. It automatically provides guidance in the proper use of the formulas and helps you keep an eye on what is truly important in physics - its ideas.

So we postulate rule number two for success in physics: **KNOW THE STORY BEHIND THE FORMULAS!**

1.3 UNITS IN PHYSICS

In physics you will be introduced to a host of units, the vast majority of which you will encounter for the first time. New units emerge from virtually every chapter and topic in physics. Before long you will meet the Joule, Newton, Watt, Coulomb, Ohm, Volt, Ampere, Tesla, and so on and on. It will be necessary for you to know the definitions of these units. The Newton (N), for example, is defined as a kilogram · meter per second squared ($kg \cdot m/sec^2$); the Watt (W) is defined as a Joule per second (J/s), which is equivalent to a Newton · meter per second (N·m/s), which in turn is identical to a kilogram · meter squared per second cubed ($kg \cdot m^2/s^3$), and so on. An elaborate structure of units based on units is developed in physics, one that appears at first glance as a formidable tower of details to absorb.

What is the best way to handle this aspect of physics? Not by turning the course into another list of items to memorize! What was said for formulas applies also to units. The key is to know and understand the story behind each of them.

Units are important in physics and all the sciences because science concerns itself with measureable quantities. If you measure

General Suggestions 11

the height of a telephone pole and report it to be 18 feet long, you are saying that the length of the pole is 18 times as long as one foot. Had you omitted the word *feet* and reported the length merely as 18, your statement would have no meaning at all. No one would be able to discern how tall the pole is from your report. Is it very short, such as only 18 inches tall, or is it of medium height, such as 18 yards, or is it very tall, reaching up to a height of 18 miles? Even with the word feet *in* your report, the statement provides no information to anyone who does not know how long a foot is. Only if the size of one foot is known does the report have meaning. It conveys the fact that the pole is 18 times as long as one foot.

Think of it this way. What if the foot would have been somewhat longer or shorter than it is? The length of the pole would then not be 18 anything! The pole is standing out there somewhere, reaching up to some height. There is nothing about its length that automatically designates it to be 18. It is only because we chose to make the foot as long as we did that the pole turns out to be 18 times as long. Had we chosen otherwise, the number 18 would never arise! Had we invented no standard of length at all, we wouldn't be able to attach any number to the height of the pole.

In other words, the expression "18 feet" to describe the length of the pole is entirely dependent on our chosen standard of length, in this case the foot. The foot serves as our *unit* of length, meaning it defines how long "one" is.

There are many units in physics because physics concerns itself with many different types of measureable quantities. We need a unit for force - so we create the Newton, a unit for charge - the Coulomb, a unit for electrical resistance - the Ohm, and so on. The first and foremost thing to know about a unit is the quantity for which it serves as a unit. The foot is a unit of length, the Newton a unit of force, the Coulomb a unit of charge, the Ohm a unit of resistance, and so on. Always associate units with quantities.

Units in physics are classified either as "fundamental" or "derived". A very small number of units are of the fundamental variety. Their definitions depend on no other units, but stand on their own. In this catagory are the units of length, mass and time. Throughout this book we will use the MKS (meter-kilogram-second) system of units. The unit of length in this system, the meter, is defined as the distance between two scratches made on a platinum-iridium rod that is kept at the International Bureau of Weights and Measures in Sevres, France (when the rod is at a temperature of 0 degrees Celsius). This is about the length of the ordinary meter stick common to most laboratories and classrooms.

The unit of mass in this system, the kilogram, is defined as the amount of matter contained in a platinum-iridium cylinder housed in the same bureau, near Paris. This is about the amount of matter contained in the one-kilogram cylinders commonly found in most laboratories and classrooms. The unit of time, the second, is the same as the second in widespread use today, of which there are 60 in one minute and 3600 in one hour. The hour in turn is defined as one twenty-fourth of the average time that elapses between one noon and the next, over the course of one year.

The above designations were decided upon by a commission of scientists in 1790. The platinum-iridium alloy was used because it resists corrosion.

The overwhelming majority of units, however, are of the derived variety. These units are defined in terms of other units. The unit of velocity in the MKS system, for example, is defined as the velocity of an object that travels one unit of length, the meter, in one unit of time, the second. The velocity of any object in meters per second (m/s) is determined by dividing the number of meters traveled by the number of seconds elapsed. In a similar vein, the unit of acceleration in the MKS system is defined as the acceleration of an object that gains one unit of velocity, a meter per second, in the course of one unit of time, the second. The acceleration of any object in meters per second per second (m/s/s) is determined by dividing the number of meters per second of velocity gained by the number of seconds elapsed. When meters per second are divided by seconds we arrive at meters per second squared (m/s^2).

Derived units may be defined in terms of other derived units, as demonstrated above in the case of the unit of acceleration. Ultimately, however, all derived units can be expressed as some combination of fundamental units.

The definition of every derived unit is based on the nature of the quantity for which it serves as a unit. That the meter per second is the unit of velocity is a result of the meaning of velocity. The meter per second squared is the unit of acceleration due to the meaning of acceleration. Similarly, the definition of the unit of force, the Newton, as one kilogram · meter per second squared, follows from the definition of the quantity we call force (mass times acceleration). Once you attain a clear understanding of the nature of a particular quantity the definition of its unit follows almost automatically. You may have to memorize some names, such as that the unit of force is also called a Newton. But you will have no difficulty recognizing the unit (such as the Newton) for what it is and represents (in the case of the Newton - one kilogram·meter per second squared).

Just as the key to handling formulas is not to focus on

memorizing them and their rules and restrictions, but to know the stories behind them, that is, understanding the ideas and processes whence they come, so ought units be dealt with. The key to mastering them is to associate each unit with the quantity for which it serves as a unit, and to understand the meaning of that quantity. Want to know how the unit of acceleration is defined - focus on the meaning of acceleration. Want to know how the unit of force is defined - your best approach is to ask yourself, "What is the meaning of force?"

So we postulate rule number three for success in physics: **ASSOCIATE UNITS WITH THEIR QUANTITIES!**

A helpful way to practice making these associations, and thereby remember them, is to analyze all equations for balance insofar as units are concerned. This procedure is referred to as *dimensional analysis*. Every valid equation must be balanced dimensionally, that is, the units represented on either side of the equation must be identical. The analysis is performed by expressing every quantity contained in the equation in terms of its units, then performing the operations indicated in the equation on those units. Sometimes it is necessary to reduce derived units to fundamental ones before it becomes obvious that both sides of the equation do indeed have identical units.

As an example, consider the equation for distance traveled by an accelerating object, $d = v_i t + \frac{1}{2} a t^2$. First we express all quantities in terms of their units, as such:

$$d = v_i t + \frac{1}{2} a t^2$$

$$\text{meters} = \left(\frac{\text{meters}}{\text{sec}}\right)(\text{sec}) + \left(\frac{\text{meters}}{\text{sec}^2}\right)(\text{sec})^2$$

General Suggestions 15

Notice that the number 1/2 has no units. This is due to the fact that it does not represent a measureable quantity. We label such numbers "dimensionless".

Next we perform the indicated operations in the proper order:

$$\text{meters} = \frac{\text{meters} \cdot \text{sec}}{\text{sec}} + \frac{\text{meters} \cdot \text{sec}^2}{\text{sec}^2}$$

$$\text{meters} = \text{meters} + \text{meters}$$

Since meters plus meters yield more meters, we arrive at:

$$\text{meters} = \text{meters}$$

and the equation is indeed balanced dimensionally.

Doing this repeatedly for every equation as they are introduced focuses the mind on the units involved and immensely assists the process of internalizing the association between unit and quantity. Mastering these associations is essential to success in physics.

The fundamental units described earlier (meter, kilogram, second) form what is known as the MKS system of units. Other systems have gained acceptance over time, though none is as widely used in physics today as the MKS system. Frequently it is necessary to convert a unit from one system to another. For example, a problem might present speed values in terms of kilometers per hour (km/hr) but the physics formulas to be used in solving the problem are designed to accept only meters for distance and seconds for time. So the student must translate, say, 80 km/hr into an equivalent number of meters per second. The student knows there are 1000 meters in one km and 3600 seconds in one hr. So how many m/s are equal to 80 km/hr?

The best method for performing such conversions, one that is not as error prone as other methods and that animates you to focus on the key concepts involved, is the "substitution" method. It consists of substituting new units (the desired ones) in place of the old (undesired ones) and performing the indicated algebraic operations. To convert 80 km/hr to m/s proceed as follows:

$$\frac{80 \text{ km}}{1 \text{ hr}} = \frac{(80)(1000 \text{ m})}{(1)(3600 \text{ sec})}$$

This was accomplished by replacing the word *kilometer* with *1000 meters* (its equivalent) and the word *hour* with *3600 seconds*. The next step is to perform the indicated algebraic operations. Since the 80 was to be multplied by kilometers (80 km represents a distance 80 *times* as long as one km), we now multiply the 80 by 1000 meters; Similarly we multiply 1 by 3600 seconds. We arrive at:

$$\frac{80,000 \text{ m}}{3600 \text{ sec}} = 22.2 \text{ m/sec}$$

1.4 SYMBOLS IN PHYSICS

The formulas and units that permeate physics have spawned a large number of symbols to represent the many quantities and units encountered in the subject. The majority of these symbols are of the "first letter in the word" type and are thus easy to recognize. Acceleration rate is typically symbolized by a, velocity by v, time by t, and so on.

This is not the case, however, for all the symbols. Some names of quantities begin with the same letter as others, so they cannot all be represented by "first letter in the word" type symbols. A prominent example of this is mass and momentum. The common practice is to let m represent mass, while using p to represent momentum. But p also represents power. The net result is that we

General Suggestions 17

have no choice but to make whatever effort is required to keep track of the symbols we use.

The situation is further complicated by the need for subscripts and superscripts. In collision problems, for example, we need to distinguish between the velocity of one object and another, before and after a collision. How can we represent all four velocities by the letter v and still allow (in the formula for conservation of momentum, for example) their values to be different? The common practice is to label the velocity of one object before collision v_1 and the velocity of the same object after collision v_1' (pronounced v_1-prime). The velocity of the other object before collision is then labeled v_2, and its velocity after collision v_2'. The subscripts denote the object number and the presence or absence of the prime superscript reveals whether it is the velocity before or after the collision that is being represented.

Is this cumbersome system easier to manage and less error prone than to simply label the four velocities a, b, c and d? Adopting the a, b, c, d system would eliminate the clumsy subscripts and superscripts (they are the source of many errors) but would necessitate memorizing which letter represents which velocity. Either way we must pay careful attention to the symbols we employ.

If you are disconcerted by all the strange, complicated-looking symbols used in physics, keep in mind that over time, with repeated use, you are likely to become increasingly comfortable with their presence. The best advice in this regard is to make sure you know the precise meaning of every symbol before using it and to practice using the symbols in appropriate contexts - such as in the process of solving problems.

So we postulate rule number four for success in physics: **PRACTICE THE CAREFUL USE OF SYMBOLS!**

1.5 SOLVING THE PROBLEMS

The ultimate challenge posed by physics, in the perception of many, is solving the problems that usually appear at the end of every chapter. Students never cease to be amazed at the fact that the problems frequently seem difficult if not impossible to solve even when they are certain they understand the material. And they really do understand the material; yet that does not automatically guarantee the ability to solve the problems. Why is this so? What can be done about it?

The key point is this: There are two sides to physics - understanding its ideas and using those ideas. Applying the ideas of physics in the process of solving problems can be referred to as "doing physics". Doing physics is a skill based on, but independent of, understanding physics. Understanding physics is a pre-requisite to doing physics but is not sufficient to automatically lead to the ability to do physics. To put it very plainly: One can truly understand physics, know all there is to know about it, and still not be adept at solving its problems.

An appropriate analogy comes to mind. Consider such areas of human endeavor as auto repair, surgery, plumbing and carpenting. A person who studied the most authoritative books on plumbing technique, or who attended the best medical schools where he or she absorbed great lectures on proper surgical procedure, or who earned excellent grades on all tests administered by an auto repair

General Suggestions 19

school, is still not considered to be a good surgeon, plumber, auto mechanic or carpenter until he or she has obtained what is commonly referred to as "hands on experience". Would you trust your car or, better yet, your health in the hands of someone who earned an *A* in Auto Mechanics or Surgical Technique but never actually worked on a car or successfully performed surgical procedures? Not unless you're desperate! Why is this so? Did not the candidate earn a perfect score? Because there is a world of difference between merely knowing something and having that expert touch that comes only with experience. Knowledge is not synonomous with "being able to do". This is why prospective doctors must go through an extensive period of supervised service known as "internship" before they can start practicing medicine on the public at large.

Students of physics typically learn the subject from books and teachers. After absorbing the polished presentations of a teacher and reading the concise explanations of a textbook, it is easy to be lulled into believing that the material has been mastered. And indeed it may very well have been - in the sense that the concepts are fully understood. But this does not mean that the student will experience no difficulty in solving the problems at the end of the chapter. The ability to solve those problems efficiently and smoothly, especially the more challenging ones, comes only with "hands on" experience. The student must practice solving problems in an appropriately organized manner in order that he or she becomes skillful at recognizing the procedures and tactics most likely to prove effective in solving different types of problems.

Problem solving in physics demands more than conceptual mastery. It requires decision making skills of a high order. Choices have to be made, frequently in an organized sequence. Certain questions need to be asked and answered. The problem solver is on his or her own and must make *original* decisions. (Unless the student recollects seeing the solution steps to a similar problem - then all he need do is copy-cat those steps. But then he is not solving a problem, just recollecting steps.) Even if the lectures or text materials include solution steps to sample problems, as they sometimes do, that still doesn't provide sufficient guidance in solving a problem that is structured differently. The ability to make a sequence of correct moves, those that lead to the correct result in an efficient manner, is a skill entirely independent of the quality of one's understanding of the concepts.

A common mistake students make in the study of physics, one made even by the more diligent types, is to follow up on the discovery that they can't solve certain problems by "finding out" how the troublesome problems are to be done and leave it at that. They would consult a teacher, friend or book for the correct procedure to employ in order to arrive at the correct answer. When the student then perceives that he or she understands that procedure, the problem (of not having been able to solve the problem) is deemed to have been resolved and the student is ready for the test. After all, he or she now *knows* how to solve the problems that couldn't be solved earlier.

Not so! Knowing and understanding a problem solving procedure presented by someone else means just that and no more - you understand the procedure for that type of problem. But knowing and understanding is not synonomous with *being able to do.* Knowing and understanding does not mean that you can solve problems on the same subject *on your own,* when *your* decision

General Suggestions 21

making skills will be in demand. This is especially so for problems that are structured differently from those whose solution steps are supposedly understood.

If you find that you are encountering difficulty solving certain problems on a particular subject, by all means consult any available resource to ascertain how the problems are to be solved. But don't stop there! Go back and try, on your own, a variety of other problems on the same subject. You are not ready for the test until you have demonstrated to yourself that you can solve a wide variety of problems *on your own*.

The practice of problem solving, to be useful and productive, must be based on a clear understanding and complete knowledge of the concepts, definitions and formulas contained in a particular topic and the problems must be addressed in a logical and systematic manner. It is imperative that sound procedural habits be developed and nurtured over time. Below are some suggested steps to take in the course of seeking the solution to a problem. We will repeatedly employ and demonstrate the effectiveness of these steps as we tackle problem after problem throughout the remaining chapters of this book.

A good first step to take in solving almost any physics problem is to make a sketch of the problem. This need not consist of elaborate and time consuming art work, only a bare-bones outline of the facts presented in the problem need be drawn. It is important, however, that the sketch incorporate facts not explicitly presented in the problem but which can be inferred from your knowledge of physics. It is also helpful to employ certain coded symbols, such as the single arrow ($\longrightarrow$) to represent motion and the double arrow ($\Longrightarrow$) to represent force.

As an example consider the following problem. A 4000 N car coasts at the rate of 80 m/s. It is then subjected to a force of friction of 50 N. How long does it take the car to be brought to a stop and how far does it travel during that time? A sketch that helps focus the mind on the relevant facts of this problem might look like this:

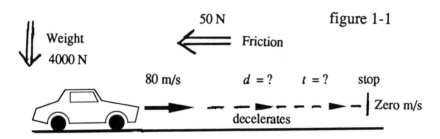

figure 1-1

Notice the single arrow directed to the right representing the car's velocity, the double arrow directed leftward representing the force of friction (which is always directed opposite to motion) and the double arrow directed downward representing the force known as weight. A clear and helpful sketch does not blur the distinction between different quantities, such as force and velocity, by assigning the same symbol to both. Notice also that although the problem does not explicitly state in which direction the forces of friction and weight are acting, constructing the sketch animates you to dig out and focus upon facts such as these that are submerged between the words of the problem. A good sketch can therefore be a powerful tool in the hands of the problem solver.

A good next step to take is to classify the story depicted in the problem. This helps draw your attention to the principles, ideas, definitions, and formulas that are relevant to the topic and are therefore most likely to have a bearing on the problem at hand. If it's a motion problem, ask yourself: what catagory of motion - uniform,

General Suggestions 23

accelerated or decelerated - is involved? If forces exist, are they gravitational, frictional, electrical, elastic, magnetic or nuclear forces? Certain features of a problem stand out as benchmarks that serve to classify the problem as belonging to a particular domain of physics. In the above example, we have decelerated motion under the influence of one force - friction (the engine is not acting on the car since the car is said to be "coasting").

The next step is to identify and list, certainly in your mind but better yet on paper, all the known values in the problem and all those being sought. This may involve a thorough search and examination of the words of the problem, because frequently more data is provided in a problem than at first meets the eye. It is therefore important to spend time carefully studying the problem in order that all the data provided, those obvious and those submerged, are noticed and identified. To say, for example, "an object is released and falls" is to imply that the starting speed of its downward trip was zero m/s. In the example given above we are provided with v-initial (80 m/s), v-final (zero m/s), weight (4000 N) and a single (and therefore *net*) force of 50 N directed opposite to motion. We seek distance and time.

Now comes the most challenging part of the process - discovering how to go from the known values to those being sought. Your mind scans the concepts, definitions, relationships, formulas and principles pertaining to the story depicted in the problem to see if any one of them, or some combination of them, enables you to link the known quantities to those being sought. The process is particularly difficult when the linkage is not provided by any one item (formula, definition, law, etc.) but by a combination of items. Finding the right combination may require much scanning -

which is why listing and sketching on paper is frequently very helpful. Your mind may travel through many blind alleys before setting upon the path that meets with success. Success in this case consists of establishing a link that enables you to determine an unknown value from a known value.

To appreciate what this mental scanning process is like, let us take a trip through a possible line of reasoning in the case of the example above. Chances are your attention would be drawn early to the formulas that have an isolated d (for distance) on one side of the equal sign (isolated in the sense that no other symbols appear on the same side), since d is one of the quantities being sought. Since you classified the problem as one involving decelerated motion, your focus would be helpfully confined to formulas within that realm. You would not be erroneously lulled into employing the relationship $d = vt,$ for you are well aware that that formula is applicable only to situations of uniform velocity. Instead your attention would gravitate to $d = \bar{v}t$ or to $d = v_i t + \frac{1}{2}at^2$, both of which apply to accelerated or decelerated motion. If you turn your attention to $d = \bar{v}t$ it immediately becomes clear that other pieces of the puzzle, namely $\bar{v}$ (average speed) and t (elapsed time), are missing. This brings you to a fork in the road. A decision must be made. Either you drop $d = \bar{v}t$ as useless due to the lack of necessary information and turn to other possiblities (such as the other formula with an isolated d) or you stay the course and proceed to search for the missing data - average speed and time. As you continue to scan the pertinent tools for linkages your mind comes across the formula $\bar{v} = (v_i + v_f)/2$ and you notice that all the information needed to put it to productive use is at hand. Since $v_f = 0$ m/s (the car comes "to a stop") and v_i is 80 m/s (given), you proceed as follows:

General Suggestions

$$\bar{v} = \frac{v_i + v_f}{2} = \frac{80 + 0}{2} = 40 \text{ m/s}$$

Now you know $\bar{v}$, the average speed, and all that is needed in order to extract the distance from $d = \bar{v}t$ is the elapsed time, t. So your mind continues to scan the available tools (formulas, concepts, definitions, principles, etc.) for one that will yield a value for t. You notice that the formula $v_f = v_i + at$ would lead to knowledge of t if only a (the acceleration rate) were known (recall that v_f and v_i are already known). Can a be found somehow? Perhaps from $F = ma$. We know F (the net force - in this case the only force, that of the 50 N frictional force) but not m, the mass. Can m somehow be ascertained? As your mind continues to scan the available tools you notice that m can be obtained from $W = mg$, since W (the weight) is given (4000 N) and g is a known constant (9.8 m/s²). You proceed as follows:

$W = mg$
$4000 = m\,(10)$ (approximating 9.8 as 10)
$m = 400$ kg

Proceeding now to retrace your steps through the maze, you continue:

$F = ma$
$50 = (400)\,(a)$
$a = .125$ m/s²
$v_f = v_i + at$
$0 = 80 - (.125)(t)$ (a is negative since the car is decelerating)
$80 = (.125)\,(t)$
$t = 640$ sec

Now, back to the starting point:

$d = \bar{v} t$

$d = (40)(640)$

$d = 25{,}600$ meters

We can illustrate the steps taken in solving this problem in the form of a tree, as such:

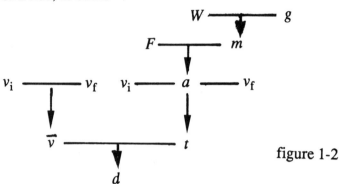

figure 1-2

Could the problem have been solved differently? Yes. There frequently is more than one path to the solution of a physics problem. For example, after determining that a is .125 m/s^2, one could turn to the formula $v_f^2 - v_i^2 = 2ad$ and proceed as follows:

$v_f^2 - v_i^2 = 2\,a\,d$

$0 - 80^2 = 2(-.125)\,d$

$6400 = .25\,d$

$d = 25{,}600$ meters

This would have saved a step - the one taken to find average speed. But we are not obligated to find the shortest, quickest route to the answer! As long as we don't take a wrong turn, by making a move that is incorrect, we ought to be satisfied. We have correctly solved the problem.

A key point that emerges loudly and clearly from the above demonstration of the "scanning for linkages" process is that it is very much helped along by the suggested preliminary steps:

General Suggestions 27

Identifying all the quantities given and those being sought (this lets you keep track of what is to be linked to what), classifying the story of the problem (this lets you confine the scanning to the pertinent tools, thereby limiting the field and precluding erroneous moves) and drawing the sketch (this helps in the execution of the other steps). The sequence of steps in solving a problem should be: SKETCH, CLASSIFY, IDENTIFY, SCAN AND LINK.

Problem solving can be an enjoyable and successful endeavor if it is approached in a systematic, logical manner. As you develop the habit of solving problems in an orderly series of steps you will become better and more skillful at it. In the remaining chapters of this book we aim to inculcate these good problem solving habits by demonstrating and practicing them repeatedly as we move through the various topics covered in a typical first year physics course.

Rule number five for success in physics is: **DEVELOP AND PRACTICE GOOD PROBLEM SOLVING HABITS!**

1.6   CONCEPT OF PROPORTIONALITY

Establishing relationships between quantities is an important aspect of physics, one that is encountered quite often. After all, the business of physics is to understand how the universe operates and what its rules are. Those rules, sometimes referred to as "laws of nature", are typically expressed in the form of relationships between measureable quantities. A formula in essence is such a relationship expressed mathematically.

A quantity is said to depend on another if a change in the value of one quantity leads to a change in the value of the other. For example, your height depends in some manner on your age. As your age changes your height is likely to change with it. Analogously, the force of gravity between objects is dependent on the distance between them, the speed of a falling object depends on how much time it has been falling, the size of an image formed by a lens depends, among other things, on the distance between the object and the lens, and so on. Quantities that can change in value are referred to as *variables*.

The manner in which a change in one variable leads to a change in another is known as *variation*. Many different types of variation are encountered in physics. A relationship of *direct variation* is said to exist between quantities if multiplying the value of one of them by some number leads to multiplication of the corresponding value of the other quantity by the same number. Such a relationship exists, for example, between the net force acting on an object and the object's acceleration rate (so long as the object's mass remains unchanged). If a twice as strong net force is exerted on an object (multiplication by 2), the acceleration is doubled. If a three times as strong net force is exerted (multiplication by 3), the acceleration is tripled. If the net force is reduced to one half its original strength, the acceleration becomes one half the amount it used to be. And so on.

To be more specific, if a force of 6 Newtons acting on a particular rock produces an acceleration of 15 m/s^2, then a force of 12 N will accelerate the same rock at the rate of 30 m/s^2, a force of 18 N will yield an acceleration of 45 m/s^2 and a force of 3 N will cause the rock to accelerate at the rate of 7.5 m/s^2.

When a relationship of direct variation exists between two quantities, we say each quantity is *proportional to* the other. For

General Suggestions

example, the acceleration of an object is proportional to the net force acting on it. We write this mathematically as follows:

$$a \; \alpha \; F$$

where the symbol α represents the phrase "is proportional to". More generally, if any two variables x and y are related to each other by direct variation, we write:

$$y \; \alpha \; x$$

It can be shown mathematically that whenever two quantities, x and y, are related to each other by direct variation, their values are linked together by an equation of the form

$$y = kx$$

where k is some fixed number. The number k is known as the *constant of proportionality*. No other form of an equation can satisfy the direct variation relationship.

As an example, let us choose the value 3 for k and list some x and y values that fit into the equation $y = 3x$. This is done in table 1-1.

X	Y
1	3
2	6
3	9
4	12
5	15
6	18
7	21

TABLE 1-1

Notice that as x is doubled from 2 to 4, y is also doubled - from 6 to 12. As x is tripled fron 1 to 3, y is also tripled - from 3 to 9. As x is multiplied by 1.5, from 4 to 6, y is also multiplied by 1.5 - from 12 to 18. Any multiplicative change in x leads to the same multiplicative change in y.

We would have obtained the same pattern if k had been chosen to be any other number, so long as the equation linking x to y is of the form $y = kx$. If we try this, however, with an equation of any other form, it will not work. For example, let us try $y = 3x + 1$. Pairs of (x, y) values that satisfy this equation appear in table 1-2.

X	Y
1	4
2	7
3	10
4	13
5	16
6	19

TABLE 1-2

As x is doubled from 2 to 4, y goes from 7 to 13 and is not doubled. As x is tripled from 2 to 6, y goes from 7 to 19 and is not tripled. Clearly this is not a direct variation relationship.

If pairs of numbers linked together by direct variation are plotted on a graph of cartesian coordinates, the pattern forms a straight line.

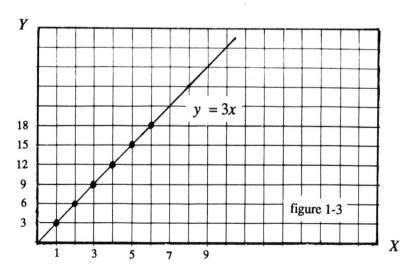

figure 1-3

General Suggestions

The y-intercept of the line is zero, the line intersect the origin, and the slope of the line is equal to the constant of proportionality. In figure 1-3 we plot some of the values linked together by the equation $y = 3x$. Clearly, the points representing the linked values form a straight line that intersects the origin. That the slope of the line is equal to 3 - the constant of proportionality in this example - can be determined by choosing any two points on the line, sketching the rise (Δy) and the run (Δx) between them and finding the ratio $\Delta y/\Delta x$ (the definition of slope). This is illustrated in figure 1-4.

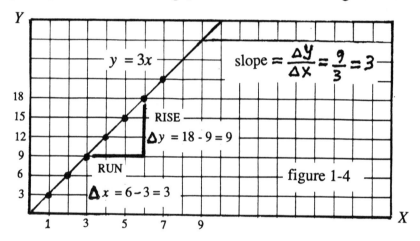

Another type of relationship between variable quantities is known as *inverse variation*. Inverse variation exists when multiplying the value of one quantity by a number leads to multiplication of the corresponding value of the other quantity by the inverse of the number. If one quantity is multiplied by 2, the corresponding value of the other is multiplied by 1/2. If one quantity is multiplied by 3/2, the other is multiplied by 2/3. And so on. Such a relationship exists, for example, between the force exerted by one current on another and the distance between the currents.

It can be shown mathematically that when two quantities x and y are related to each other by inverse variation, their values are linked together by an equation of the form

$$y = \frac{k}{x}$$

where k is some fixed number. No other form of an equation can satisfy the inverse variation relationship.

To see how this is so, let us choose the number 4 for k and list some pairs of values that satisfy the equation $y = 4/x$. This is done in table 1-3.

X	Y
1	4
2	2
3	4/3
4	1
5	4/5

TABLE 1-3

We see that as x is doubled from 1 to 2, y is halved from 4 to 2. As x goes from 2 to 3 (multiplication by 3/2), y goes from 2 to 4/3 (multiplication by 2/3). Clearly these quantities are linked by an inverse variation relationship.

Since the equation $y = k/x$ can also be written as $y = k(1/x)$ we can express the inverse variation relationship by saying "y is proportional to the inverse of x". This is expressed mathematically as follows:

$$y \propto \frac{1}{x}$$

In figure 1-5 we plot the linked pairs of numbers listed in the example above on a graph of cartesian coordinates. This type of curvature is typical of the inverse variation relationship.

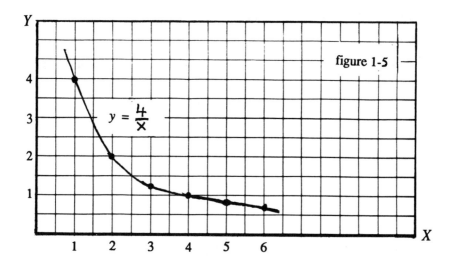

figure 1-5

Another type of relationship between variable quantities, one that makes a frequent appearance in physics, is known as *inverse square variation.* Inverse square variation exists when multiplying the value of one quantity by a number leads to multiplication of the corresponding value of the other quantity by the square of the inverse of the number. If one quantity is multiplied by 2, the corresponding value of the other is multiplied by the square of 1/2, or 1/4. If one quantity is multiplied by 3/2, the other is multiplied by the square of 2/3, or 4/9. And so on. Such a relationship exists, for example, between the force of gravity and the distance between two objects.

It can be shown mathematically that if two quantities x and y are related to each other by inverse square variation, their values are linked together by an equation of the form

$$y = \frac{k}{x^2}$$

where k is some fixed number. No other form of an equation can satisfy the inverse square relationship.

As an example, let us choose the number 5 for k and list some pairs of numbers linked together by the equation $y = 5/x^2$. This is done in table 1-4. We see that as x goes from 1 to 3 (multiplication

X	Y
1	5
2	5/4
3	5/9
4	5/16
5	5/25

TABLE 1-4

by 3), y goes from 5 to 5/9 (multiplication by 1/9). One-ninth is the square of one-third, which is the inverse of three. As x goes from 2 to 4 (multiplication by 2), y goes from 5/4 to 5/16 (multiplication by 1/4). One-fourth is the square of one-half, which is the inverse of two. Clearly, the equation produces pairs of numbers linked by an inverse square relationship.

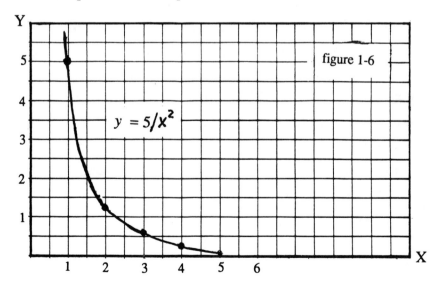

General Suggestions 35

Since the equation $y = k/x^2$ can also be written as $y = k(1/x^2)$ we can express the inverse square relationship by saying "y is proportional to the inverse of the square of x". This is expressed mathematically as follows:

$$y \: \alpha \: \frac{1}{x^2}$$

In figure 1-6 we plot the linked pairs of numbers listed in the example above on a graph of cartesian coordinates. This type of curvature is typical of the inverse square variation relationship. It is roughly similar in appearance to the graph of inverse variation, but the precise shape of the curvature is different.

2

CHAPTER TWO

MOTION

2.1 VOCABULARY AND FORMULAS

WORD	SYMBOL	UNIT
Distance	d (or s)	meter, m
Time	t	second, s
Velocity, speed	v	m/s
Average velocity (or speed)	$\bar{v}$	m/s
Change in distance	Δd	meter, m
Change in time	Δv	second, s
Acceleration	a	m/s²
Initial velocity	v_i (or v_0)	m/s
Final velocity	v_f (or v)	m/s
Free fall acceleration	g	9.8 m/s²

FORMULAS

a. $\bar{v} = d_T / t$ b. $v = \Delta d / \Delta t$ c. $d = vt$ d. $a = \Delta v / \Delta t$

e. $v_f = v_i + at$ f. $d = v_i t + at^2 / 2$ g. $v_f^2 - v_i^2 = 2ad$

h. $\bar{v} = (v_i + v_f) / 2$

Motion

2.2   THINGS TO KNOW

(1) The symbol for distance in the above formulas, d (or s in some texts), represents the distance between where an object is at any time t on a running clock and the point where the object was when the clock was started (t set to zero). The formulas are designed for motion along a straight line. They can be used through any time t after the clock was started so long as no constant appearing in a formula has changed value.

For example, the formula $d = vt$ cannot be used in the case of an object that changed direction or speed before the clock reached the value of time being plugged into the formula. This is so because v is assumed constant in this formula (a change in direction implies a change in the sign of the velocity from positive to negative or vice-versa). Likewise, the symbol a is assumed constant in formulas (e), (f) and (g).

(2) The quantity "distance traveled" between when the clock was started ($t = 0$) and any time t later is *not* necessarily the same as the distance between two points, d, described in (1) above. It *is* the same as d only if the object moved in one direction, in a straight line, the entire time.

(3) If you must compare two points in time neither of which is zero, things get a bit complicated. Say you're looking for the distance traveled (in a straight line, in one direction) between $t = 4$ sec and $t = 7$ sec. The formulas are not designed for this; they work only for starting times of $t = 0$. But all is not lost. You can proceed as follows. First find the distance between where the object was at $t = 0$ and where it is at $t = 7$ by plugging $t = 7$ into the appropriate distance formula. Then find the distance between where the object

was at $t = 0$ and $t = 4$ by plugging $t = 4$ into the formula. Subtracting the latter distance from the former yields the distance traveled between $t = 4$ and $t = 7$, provided that the object traveled in a straight line, in one direction.

Alternatively, we can make believe that another clock was started ($t = 0$) when our clock read $t = 4$. This new clock reads $t = 3$ when our clock arrives at $t = 7$. We now find d on the new clock, from $t = 0$ to $t = 3$, using the appropriate formula. This is the distance we are looking for. However, this approach may compel us to search for v-initial, the speed of the object at the instant the new clock is started.

(4) Average velocity, $\bar{v}$, cannot as a rule be obtained by finding the average of the two or more velocity values. Instead, we find the total distance traveled, d_T, and divide by the total elapsed time, t, as provided by formula (a). Formula (h) for average velocity is applicable only to situations of uniform acceleration or deceleration.

(5) Whenever an object experiences uniform deceleration, its acceleration rate, a, is assigned a negative value. This follows from the definition of a, which is, $a = \Delta v / \Delta t$. As time continues to flow and t increases, Δt is positive. At the same time the object is decelerating, its velocity, v, decreases and Δv is negative. The ratio $\Delta v / \Delta t$ is therefore also negative.

(6) The symbols that appear in a formula are related to each other. For example, in the formula $v_f = v_i + at$ the symbols v_i and v_f refer to the starting and ending velocities of the particular time interval, t, to which the formula is applied. If the formula is used for the first 10 seconds of a trip, v_i is the velocity at the beginning of that 10 second interval, when the clock must be set at $t = 0$, v_f is

Motion

the velocity at the end of that same 10 second interval, when the clock reads $t=10$, and t must be *ten*. No other velocities, at times other than $t = 0$ and $t =10$, are permitted as plug-ins for v_i and v_f, respectively, once the 10 second interval has been decided upon.

(7) Formulas (b) and (c) may be used only in cases of uniform velocity. Formulas (d), (e), (f), (g) and (h) are designed for situations of uniform acceleration or uniform deceleration. Formula (a) can be used under any conditions, so long as it is used correctly.

(8) We have the right to "start the clock" whenever we wish. In other words, t can be set equal to zero at any point in a developing story. But we must be aware of our choice, since everything else is based on it. Once a point in time is designated $t = 0$ the object's velocity at that instant becomes v_i, all t values represent time elapsed from that instant, and d represents the distance from the point where the object was at that instant to where it is some time, t, later.

(9) Acceleration is not the same as gain (or loss) of velocity. Instead, acceleration, a, is gain (or loss) of velocity *per second*. There's a big difference between the two concepts. The unit of gain (or loss) of velocity, Δv, is the m/s, whereas the unit of acceleration, a, is the m/s/s or m/s^2.

(10) The accepted value of 9.8 m/s^2 for the acceleration of free fall is applicable only on earth, near sea level.

2.3 SOLVING THE PROBLEMS

(1) A car starts from rest and accelerates uniformly. It passes the 30-meter mark in 6 seconds. (a) What is the car's average velocity?

(b) What is its velocity as it crosses the 30-meter mark? (c) What is the acceleration rate of the car? (d) If the car continues to accelerate at the same rate, where will it be when 20 seconds have elapsed?

SKETCH

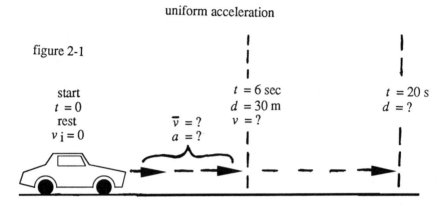

figure 2-1

IDENTIFY AND CLASSIFY

We can use the above sketch as a launching pad to help us classify and identify what is known about the problem and what we are looking for. We know the car is accelerating uniformly throughout a 20 second time interval, and we know its velocity at $t = 0$ and its distance from the starting point at $t = 6$ seconds. We are looking for its average speed during the first six second interval, the acceleration rate during those same six seconds, the velocity at the instant the clock reads $6\ sec$ and the distance from the starting point at the instant the clock reads $20\ sec$. We notice that some data apply to particular instants in time, while others are applicable over intervals of time.

LINK UNKNOWN TO GIVEN

(a) $\bar{v} = d_T/t$ (applied to first six seconds)

$\bar{v} = 30/6 = 5$ m/s

Motion

(b) $v_f = v_i + at$ (first six seconds)

$v_f = 0 + a(6)$

$v_f = a(6)$

We can't solve for v_f this way since we don't know a. Perhaps we can determine a from

$a = \Delta v / \Delta t = \Delta v / 6$

Another obstacle! We can't solve for a without knowing Δv. So we take another approach to find the desired v- final.

$\bar{v} = (v_i + v_f)/2$ (first six-second interval)

$5 = (0 + v_f)/2$

$v_f = 10$ m/s

(c) $a = \Delta v / \Delta t = (10 - 0)/(6 - 0) = 10/6 = 1.67$ m/s^2

(d) $d = v_i t + at^2/2$ (applied to twenty-second interval)

$d = 0 + (1.67)(20)^2/2$

$d = 333.33$ m

Or, taking a different approach, we notice that v- final for the first six-second time interval, namely 10 m/s, is also v- initial for the last fourteen-second time interval. We then proceed as follows:

$d = (10)(14) + (1.67)(14)^2/2$ (last fourteen seconds)

$d = 140 + 163.33$

$d = 303.33$ m

This is the distance traveled during the last 14 seconds. To this must be added the 30 meters traveled during the first 6 seconds to obtain the distance from the starting point to where the car is when the clock strikes *20 sec*. Adding 303.33 and 30 yields 333.33 meters - the same answer we obtained the other way.

(2) An airplane approaches a landing strip with a velocity of 300

km/hr and brakes to a stop in 15 seconds. (a) What is the average deceleration rate of the plane? (b) How long must the landing strip be?

SKETCH AND IDENTIFY

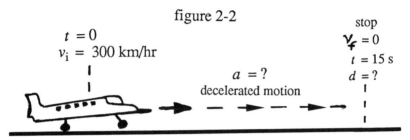

figure 2-2

LINK UNKNOWN TO GIVEN

(a) Average acceleration can take the place of uniform acceleration, a, for any particular time interval - so long as we keep in mind that, since the acceleration is not necessarily uniform, the average acceleration for any part of the time interval may differ from that for any other part.

$v_f = v_i + at$ (entire fifteen second interval)

$0 = 83.33 + a\,(15)$ (converting 300 km/hr to 83.33 m/s)

$15a = -83.33$

$a = -5.55 \text{ m/s}^2$

(b) The landing strip must be at least as long as the "braking distance".

$d = v_i t + at^2/2$ (entire fifteen second interval)

$d = (83.33)(15) + (-5.55)(15)^2/2$

$d = 625 \text{ m}$

OR

$\bar{v} = (v_i + v_f)/2$ (entire fifteen second interval)

$\bar{v} = (83.33 + 0)/2 = 41.66 \text{ m/s}$

$d_T = \bar{v} t = (41.66)(15) = 625 \text{ m}$

Motion

(3) An object is released and falls freely. (a) What distance does it travel during the first second? (b) What distance does it travel during the second second? (c) What is its velocity after two seconds? (d) How far has it fallen by the time its velocity is 80 m/s? (e) How long does it take for its velocity to reach 80 m/s?

SKETCH AND IDENTIFY

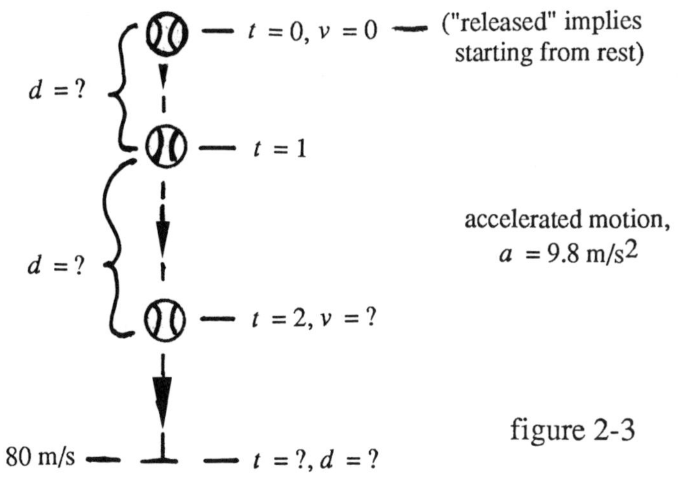

figure 2-3

("released" implies starting from rest)

accelerated motion, $a = 9.8 \text{ m/s}^2$

LINK UNKNOWN TO GIVEN

(a) Let us not confuse acceleration rate (in this case 9.8 m/s^2) with velocity. The object does *not* travel 9.8 meters during the first second. Instead we proceed as follows:

$d = v_i t + at^2/2$ (applied to first second)

$d = 0 + (9.8)(1)^2/2 = 4.9 \text{ m}$

(b) $d = v_i t + at^2/2$ (first two second interval)

$d = 0 + (9.8)(2)^2/2 = 19.6 \text{ m}$

This is the total distance traveled during the first two seconds. But the problem asks for the distance traveled during the second second. So we subtract 4.9 from 19.6 and arrive at 14.7 m.

Alternately, we could apply the distance formula to the second

second by setting $t = 0$ at the beginning of that second. But this means that v-initial (the velocity at $t = 0$) is no longer zero.

$d = v_i t + at^2/2$ (applied to second second)

$d = v_i (1) + (9.8)(1)^2/2$ ($t = 1$ at the end of the second second)

We notice that v-initial for the second second is the same as v-final for the first second. This must be so since the end of the first second coincides with the beginning of the second second. V-final for the first second can be found via:

$v_f = v_i + at$ (first second, $t = 0$ when object is released)

$v_f = 0 + (9.8)(1) = 9.8$ m/s

V-initial for the second second therefore is 9.8 m/s.

$d = v_i t + at^2/2$

$d = (9.8)(1) + (9.8)(1)^2/2$ (back to second second)

$d = 14.7$ m

(c) $v_f = v_i + at$ (first two seconds)

$v_f = 0 + (9.8)(2)$

$v_f = 19.6$ m/s (speed at end of first two seconds, $t = 2$)

(d) $v_f^2 - v_i^2 = 2ad$ (from release to when speed is 80 m/s)

$80^2 - 0 = 2(9.8)d$

$d = 327$ m

(e) $v_f = v_i + at$ (from release to when speed is 80 m/s)

$80 = 0 + 9.8 t$

$t = 8.16$ sec

(4) A car travels at the rate of 20 m/s for 50 seconds, then travels at the rate of 30 m/s for 40 seconds. What is the average speed for the entire trip?

Motion

SKETCH AND IDENTIFY figure 2-4

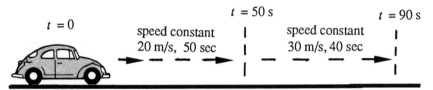

LINK UNKNOWN TO GIVEN

Average speed, $\bar{v}$, cannot generally be found by "averaging" speeds (Things to Know, item 4). Instead, the formula $\bar{v} = d_T/t$ must be used, where d_T is the total distance traveled and t is the total elapsed time. To find the total distance we first determine the distance traveled in each part of the problem, then add these together. Since each part of this problem consists of uniform velocity, we may write:

$d_1 = v_1 t$ (first fifty-second interval)

$d_1 = (20)(50) = 1000$ m

$d_2 = v_2 t$ (applied to last forty seconds, with t set to zero at the beginning of these forty seconds)

$d_2 = (30)(40)$ ($t = 40$ at the end of these forty seconds)

$d_2 = 1200$ m

The total distance, d_T, for the entire trip therefore is 1000 + 1200, or 2200 m. The total time elapsed is 90 seconds. Thus,

$\bar{v} = d_T/t$ (entire ninety seconds)

$\bar{v} = 2200/90 = 24.44$ m/s

(5) At the instant a traffic light turns green a waiting car takes off with a uniform acceleration rate of 8 m/s². At exactly the same instant another car, traveling at the constant rate of 20 m/s, passes the first car in another lane. (a) How long does it take the first car to

overtake the second? (b) How far does the first car travel before overtaking the second? (c) How fast is the first car traveling at the moment it overtakes the second car?

SKETCH AND IDENTIFY

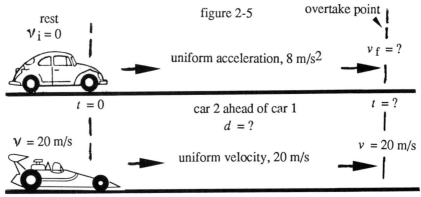

figure 2-5

Certain key words should be noticed and interpreted meaningfully as we read the problem. These facts ought then be incorporated into the sketch to help put all the information into perspective. "A waiting car" implies that it starts from rest and the car's v- initial is zero. "Passes the first car" implies that at that instant (when the traffic light turns green) both cars are at the same point on the road. We set that instant to be our starting point in time, with $t = 0$. "Overtakes the second car" implies that at some later instant both cars are once again at the same point on the road (not the same as the first point) and that both have traveled, by that time, equal distances from the starting point (the position of the traffic light).

LINK UNKNOWN TO GIVEN

(a) This problem cannot be solved by linking the data pertaining to one car only. Any formula or combination of formulas applied to one car only will contain too many unknowns to be solved. The key

Motion

to solving this problem is to link the data for both cars together. We know that both cars cover the same distance between when $t = 0$ and the instant they are together again for the second time. Keeping in mind that the first car accelerates uniformly while the second car travels at constant velocity, their distances, d_1 and d_2, respectively, at any time t, are provided by the formulas

car one	car two
$d_1 = v_i t + at^2/2$	$d_2 = v t$
$d_1 = 0 + 8t^2/2$	$d_2 = 20 t$
$d_1 = 4t^2$	

These distances are equal to each other only at the instant car one overtakes car two; they are not equal at any other time. To avoid confusion, let us designate the reading on the clock at the "instant of overtake" as t_m (for "meeting time"). We then proceed as follows:

$d_1 = d_2$ (at the instant of overtake, t_m, only)

$4t_m^2 = 20 t_m$

$4 t_m = 20$

$t_m = 5$ sec   (cars meet five seconds after light turns green)

(b) To find the distance the first car travels to overtake the second car, we replace t in the formula for d_1 with 5 :

$d_1 = v_i t + at^2/2$ (five second interval after light turns green)

$d_1 = 0 + (8)(5)^2/2$

$d_1 = 100$ m

This should, of course, be equal to d_2 for the same time interval. To see if this is indeed the case we proceed as follows:

$d_2 = v t$ (five second interval after light turns green)

$d_2 = (20)(5) = 100$ m

(c) Since the first car accelerates uniformly, its velocity, v-final, after five seconds is provided by the relationship:

$v_f = v_i + at$ (five second interval after light turns green)

$v_f = 0 + (8)(5)$

$v_f = 40$ m/s

(6) A juggler is performing his act in a room where the ceiling is five meters above the level of his hands. He launches each ball vertically upward so that it just reaches the ceiling, timing the launchings so that each ball starts going upward just as the previous one starts coming down. At what point between the juggler's hands and the ceiling do the balls pass each other?

SKETCH AND IDENTIFY

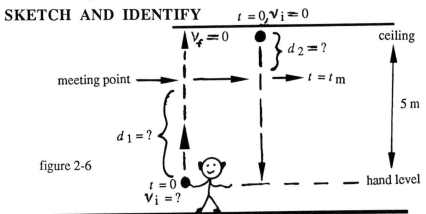

figure 2-6

We set our clock to zero at the instant ball one begins to rise, which is also the instant ball two begins to fall. When the balls pass each other the clock reads t_m ("meeting time" - analogous to what we did in problem five). The distance each ball travels by time t_m may not be equal to each other, but they must add up to 5 meters. The phrase "they just reach the ceiling" implies that the balls don't crash into the ceiling. This, in turn, means that the balls reach the

Motion

ceiling just as their velocity approaches zero (after decelerating on the way up). In other words, v-final is zero for every upward trip. All these facts are incorporated in the sketch above.

LINK UNKNOWN TO GIVEN

As in the case of problem five, this problem can be solved only by linking the data for both balls together. Keeping in mind that ball one begins its upward trip with some initial velocity and then decelerates uniformly at the rate of 9.8 m/s², the distance it has traveled by any time t after being launched is provided by the relationship:

$d_1 = v_i t + (-9.8)t^2/2$ (from $t = 0$ to any time, t, later)

Ball two, on the other hand, begins its downward trip with an initial velocity of zero, then accelerates uniformly at the rate of 9.8 m/s². The distance it travels by any time t after starting to fall is expressed by the formula:

$d_2 = 0 + (9.8)t^2/2$ (from $t = 0$ to any time, t, later)

When the clock reads t_m these distances must add up to five meters.

$d_1 + d_2 = 5$ (at time t_m only)

$v_i t_m - 4.9 t_m^2 + 4.9 t_m^2 = 5$

$v_i t_m = 5$

If we knew either t_m or v_i for each upward trip we could plug these values into the above formulas and arrive at a solution to our problem. We can attempt to find v-initial for each upward trip by noting that each such trip ends with a v-final value of zero, after covering a distance of 5 meters. We then proceed as follows:

$v_f^2 - v_i^2 = 2ad$ (applied to an entire upward trip)

$0 - v_i^2 = 2(-9.8)(5)$

$v_i^2 = 98$

$v_i = 9.9$ m/s   (every ball is launched up with this speed)

Returning now to $v_i t_m = 5$, we write:

$9.9 t_m = 5$

$t_m = .505$ sec   (time from launch to when balls meet.)

We now have all the data previously missing in our distance formulas.

$d_1 = (9.9)(.505) + (-9.8)(.505)^2/2$ (from launch to meeting point)

$d_1 = 3.75$ m

$d_2 = 0 + (9.8)(.505)^2/2$ (from ceiling to meeting point)

$d_2 = 1.25$ m

The two balls pass each other after ball one rises 3.75 meters and ball two falls 1.25 meters. These numbers do indeed add up to five meters. The balls pass each other at a point 3.75 meters above the juggler's hands.

2.4 ON YOUR OWN PROBLEMS

(Asterisk indicates solution appears in appendices.)

*(1) A car starts from rest and attains a speed of 40 km/hr in 8 seconds. (a) What is the acceleration rate of the car? (b) What is the distance traveled by the car during those 8 seconds? (c) What distance did the car travel during the third second?

*(2) A train starts from rest and accelerates at the rate of 5 m/s² for 12 seconds. It then continues at constant velocity for 100 seconds, then decelerates at the rate of 2 m/s² until it stops. What is the average speed of the train's entire trip?

*(3) (a) How far does a freely falling stone drop before it acquires a velocity of 85 m/s (if it started from rest)? (b) How long does it take the falling stone to attain this velocity?

*(4) An automobile is moving at 60 km/hr. The brakes are then applied, decelerating the car at the rate of 12 m/s^2. How far will the car travel before it is brought to a stop?

*(5) A car and a truck start from rest at the same time, with the car some distance behind the truck. The truck accelerates uniformly at the rate of 6 m/s^2 and the car accelerates uniformly at the rate of 8 m/s^2. The car catches up with the truck after the truck has traveled 200 meters. (a) How long did it take the car to catch up with the truck? (b) How far behind the truck was the car when they both started moving? (c) What is the speed of the car as it passes the truck?

*(6) Your friend falls off a cliff that is 400 meters high. You rush to the scene, arrive only five seconds after the fall and immediately dive downward (with parachute in hand, to be opened after you reach your friend). What minimum vertical (downward) velocity must the act of diving impart to you in order that you catch up with your friend before she hits the ground below?

*(7) To determine the height of a cliff you drop a stone and wait for the sound of the collision into the ground below. Assuming the speed of sound is a uniform 330 m/s, how tall is the cliff if you hear the collision 4 seconds after you drop the stone?

(8) A racehorse starts from rest and accelerates uniformly for 15 second, until it attains a speed of 60 km/hr. What distance does the horse travel during those 15 seconds?

(9) A biker moves at the uniform rate of 5 m/s for 30 seconds, then accelerates at the rate of 2 m/s^2 for 20 seconds, then moves

uniformly again for another 10 seconds. What is the average speed for the entire trip?

(10) A stone is released and falls through a distance of 100 meters. What is its speed as it arrives at the 100 meter mark?

(11) An airplane lands at one end of a runway with a speed of 120 km/hr. The runway is 3 km long. What is the minimum deceleration rate for the plane to come to a stop before it arrives at the other end of the runway?

(12) A woman on a motorcycle moving uniformly at the rate of 60 m/s passes a truck moving at 20 m/s. At the instant the motorcycle passes the truck, the truck begins to accelerate at the rate of 4 m/s^2. (a) How far behind the motorcycle is the truck 10 seconds later? (b) What is the speed of the truck as it catches up with the motorcycle?

(13) A racing car moving at the uniform rate of 200 km/hr passes you in the next lane. Eight seconds later you hear the sound of an explosion. How far from you did the racing car explode? (Assume sound travels at the uniform rate of 330 m/s.)

(14) A ball is thrown vertically upward with a velocity of 50 m/s. How many meters does it travel during the fourth second of its upward motion?

(15) The muzzle velocity of a bullet fired from a rifle whose barrel is one meter long is 600 m/s. (a) What is the average velocity of the bullet while in the barrel? (b) How much time does the bullet spend traveling through the barrel? (c) What is the acceleration rate of the bullet?

(16) A stone falls off the roof of a building. It travels through the last 70% of the distance to the ground in 4 seconds. How tall is the building?

(17) The unit of acceleration in the MKS system is the _____ .

(18) An object that starts from rest and accelerates uniformly travels how many times as much distance in the first three seconds as it does in the first second?

(19) What does a negative value for acceleration rate indicate about an object's motion?

(20) In the formula $v_f = v_i + at$, which symbols are constant and which are variable?

(21) A man on planet X discovers that any body falling from rest on his planet covers a vertical distance of 9 meters in 1.5 seconds. What is the acceleration of free fall on planet X?

(22) The barrel of a gun is held perpendicularly to the ground, pointing upward. If the muzzle velocity of a projectile launched from the gun is tripled, what happens to the height achieved by the projectile?

(23) A baseball rises vertically upward for 4 seconds. How high did it go?

(24) How long does it take the baseball in #23 to return to the ground?

(25) What is the average speed for the round trip of the baseball in #24?

3

CHAPTER THREE

MOTION GRAPHS

3.1 TYPES OF GRAPHS

UNIFORM VELOCITY GRAPHS

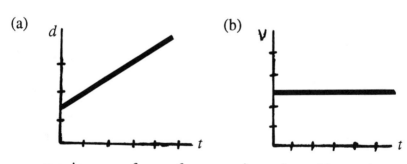

moving away from reference point, v is positive, a is zero

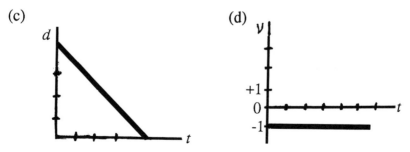

moving toward reference point, v is negative, a is zero

UNIFORM ACCELERATION GRAPHS

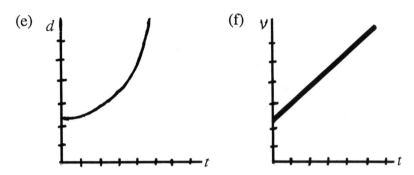

moving away from reference point, v and a positive, speeding up

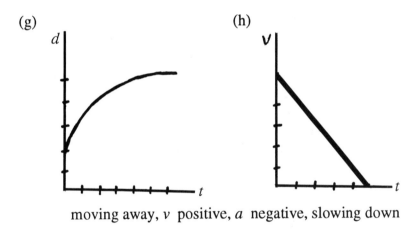

moving away, v positive, a negative, slowing down

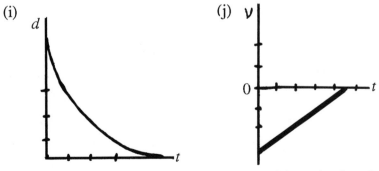

moving toward, v is negative, a is positive, slowing down

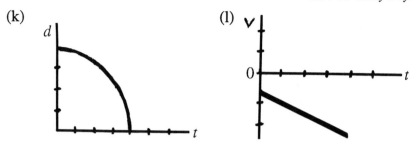

moving toward, v negative, a negative, speeding up

3.2 FORMULAS

(a) $d = d_o + vt$ **(b)** $d = d_o + v_i t + at^2/2$ **(c)** $v_f = v_i + at$
(d) $\bar{v} = \Delta d / \Delta t$ **(e)** slope of d-t graph = $\bar{v}$ (if graph is a straight line the slope is v)   **(f)** slope of v-t graph = a (straight line graphs only)   **(g)** y-intercept of d-t graph = d_o **(h)** y-intercept of v-t graph = v_i **(i)** area under v-t graph = distance traveled   **(j)** slope of tangent to d-t graph = instantaneous velocity

3.3 THINGS TO KNOW

(1) In chapter two the symbol d represented the distance between an object's position at any time t and its location at $t = 0$. In this chapter we expand our definition of d. In equations (a), (b) and (d) and in all the d-t graphs of this chapter, the letter d represents the distance between the location of an object at any time t and some chosen *reference point*. This so called reference point may or may not be the point where the object was when the clock was started ($t = 0$).

Motion Graphs

(2) The distance between the reference point and the location of the object at $t = 0$ is referred to as d-initial and is symbolized by d_o (or d_i in some texts). If the reference point *is* the location of the object at $t = 0$, then d-initial is equal to zero (and the equations of chapter two may just as well be used). The reference point, the location of the object at $t = 0$ and at all other times, t, must all lie on one and the same straight line - otherwise the equations and graphs of this chapter are not applicable (fig. 3-1).

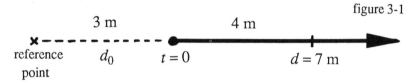

figure 3-1

(3) To understand the significance of reference points better, consider the following example. We are interested in the distance between a car and the scene of an accident at various times. The earliest observation was of a distance of 20 meters. Since we know nothing of the car's position prior to this report, this number (20 meters) becomes the distance between the car and the scene of the accident (the reference point) at $t = 0$. Later observations report smaller distances as the car approached the reference point at the rate of 4 m/s. The car's distance from the scene of the accident one second later, at $t = 1$, was 16 meters; at $t = 2$ it was 12 meters; at $t = 3$ it was 8 meters; and so on. The only equation that can incorporate these data and accurately represent the position of the car at any time t is $d = 20 - 4t$. This is the form of formula (a) above, $d = d_o + vt$, with $d_o = 20$ and $v = -4$. The equations of chapter two, none of which contain a symbol for d-initial, are just not capable of including all this information. Sometimes, as in the case of this example, we must resort to the concept of reference points

and the associated fact of a d- initial that is greater than zero.

(4) As was the case in chapter two, the symbol d in the equations and graphs of this chapter does *not* represent "distance traveled". In this chapter this is so even if the object moves in one direction all the time.

The only way to find distance traveled from the equations and graphs of this chapter is to compare the d values of two different times and subtract one from the other. In other words, we must determine Δd. And this works only if the object moves in one direction, on a straight line.

(5) The equations of this chapter can be used through any time t on a running clock so long as no constant appearing in an equation is altered during the time interval the equation is applied to. For example, the equation $d = d_o + vt$ cannot be used if the object changes direction or speed before the clock strikes the value of t being plugged into the equation. This is so because v in the equation is assumed constant.

(6) The velocity of an object moving *toward* the reference point is negative. This follows from equation (a) which can be rewritten as $v = (d - d_o)/t$. As the object approaches the reference point, the quantity d at any time t is smaller than d-initial, d_o. This makes the numerator negative, while the denominator remains positive.

(7) Don't be deceived by the looks of a graph. They are not diagrams of the motion they represent. An object may go one way, stop, then turn around and travel in the opposite direction, yet its v-t graph is a straight line (see problem one). Or a d-t graph may be curved (such as graphs (e), (g), (i) and (k) above) yet the object moves in one direction on a straight line. The only way to interpret a graph (that is, ascertain from the graph what happened to the

object) is to think about its axes, shape and slope, then compare these features to those of graphs (a) through (l) drawn above.

(8) It is possible for an object to be gaining speed (move faster and faster) yet have an acceleration rate, a, that is negative (graphs (k) and (l) above). This happens when its velocity is negative (moving toward the reference point). For example, an object whose velocity changes from -2 m/s to -5 m/s is really gaining speed in the sense that as time goes by it covers more distance per second with every passing second. But its acceleration rate is negative since -5 is, mathematically speaking, smaller than -2. Acceleration rate, a, is defined as $\Delta v/\Delta t$ (chapter two) and v, in this case, is decreasing. Thus, Δv is negative while Δt is positive, so a must be negative. Yet the object is going faster and faster - you would rather be hit by a car moving at -2 m/s than by one moving at -5 m/s!

Analogously, it is possible for an object to be slowing down yet have an acceleration rate that is positive (graphs (i) and (j) above). This also happens when v is negative and the object moves toward the reference point.

(9) To ascertain from a curved d-t graph whether the velocity is positive or negative, increasing or decreasing, one of two methods can be used. Either the particulars of graphs (e), (g), (i) and (k) are thoroughly memorized and the facts recalled as needed. Or the following more satisfying and less error prone method is employed. Two or more small *connecting lines* are drawn on the graph, each connecting two points that are relatively near each other (as illustrated in figure 3-2). Each of these lines has a slope, much as any straight line does. The slope of each connecting line is equal to the ratio $\Delta d/\Delta t$ for the small time interval between the two points it connects. As such, each of these slopes represents the average

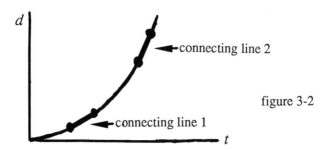

figure 3-2

velocity of the object during its respective time interval (formula (d) above). By inspecting and comparing these slopes we can ascertain whether the object's velocity is positive or negative and whether the velocity is increasing or decreasing. To do so you must, of course, remember (from algebra) that a line that leans upward-rightward (as such: /) has a positive slope, and a line that leans upward-leftward (as such: \) has a negative slope. Also, a horizontal line (—) has a slope equal to zero. You should also remember that the more vertical a line is, the greater is the absolute value of its slope and therefore the greater is the absolute value of the object's velocity.

In the graph of figure 3-2 both connecting lines lean upward-rightward, so the slope and the object's velocity must be positive throughout and the object must be moving away from the reference point. Connecting line *two* is more vertical (steeper) than connecting line *one*, so the object's velocity is increasing. Since the velocity is positive and increasing the acceleration rate is also positive.

(10) If two points on a *d-t* graph connected by a *connecting line* are very near each other, the slope of the connecting line is just about equal to the slope of the tangent drawn to the graph at either point. This approximate equality continues to improve as the points to be connected are chosen closer and closer to each other. Thus the slope of a tangent is just about equal to the average velocity of an

Motion Graphs

object during the small time interval, Δt, immediately before or after the point in time (on the graph) to which the tangent is drawn. Since this statement becomes increasingly correct as Δt is made smaller and smaller (the effect of which is to move the two chosen points closer together), we say that the slope of a tangent is the *instantaneous velocity* of the object at the point in time to which the tangent is attached. This is illustrated in figure 3-3.

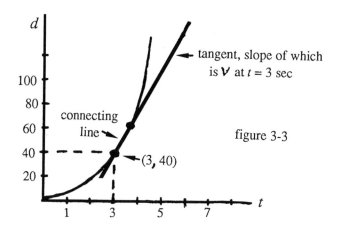

figure 3-3

3.4 SOLVING THE PROBLEMS

(1) An object is launched vertically upward with a starting velocity of 98 m/s. It rises to a certain height, then falls back to its starting point. Plot the *d-t* and *v-t* graphs for the entire trip with the ground as the reference point. Ignore the effects of air resistance.

SKETCH AND IDENTIFY

It is assumed that the problem occurs on earth, at sea level. Since the effects of air resistance are to be ignored, the object is in a state of free fall once it is launched. This means that it decelerates on the way up and accelerates on the way down at the rate of 9.8 m/s^2.

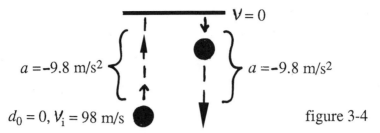

figure 3-4

Since the problem asks for a "plot", not merely a "sketch" (in which case a rough graph would have sufficed), we must determine d and v values for various times, t, via the appropriate formulas. We can apply the formulas $d = d_o + v_i t + at^2/2$ and $v_f = v_i + at$ to the entire round trip, without splitting the problem into parts, since no assumed constant in these equations is altered during the round trip (Things to Know, item five). V- final, d and t are variables, whose value is meant to keep changing or, at least, be allowed to change with the passage of time. V- initial and d- initial cannot possibly change once the object is launched. So only a remains to be considered. Well, on the way up the velocity values are positive (moving away from the specified reference point - the ground) and decreasing; on the way down the velocity values are negative (moving toward the reference point) and increasing in absolute value. This means that the acceleration rate, a, is negative (9.8 m/s^2) for both the upward and downward trips. So all assumed constants, including a, remain unchanged throughout the round trip.

LINK UNKNOWN TO GIVEN

$v_f = v_i + at$ $\qquad\qquad d = d_o + v_i t + at^2/2$

$v_f = 98 - 9.8t$ $\qquad\qquad d = 0 + 98t + (-9.8)t^2/2$

$\qquad\qquad\qquad\qquad\quad d = 98t - 4.9t^2$

TABLE 3-1

t	v	d
0	98.0	0.0
1	88.2	93.1
2	78.4	176.4
3	68.6	249.9
4	58.8	313.6
5	49.0	367.5
6	39.2	411.6
7	29.4	445.9
8	19.6	470.4
9	9.8	485.1
10	0.0	490.0
11	-9.8	485.1
12	-19.6	470.4
13	-29.4	445.9
14	-39.2	411.6
15	-49.0	367.5
16	-58.8	313.6
17	-68.6	249.9
18	-78.4	176.4
19	-88.2	93.1
20	-98.0	0.0

When the above *v-t* and *d-t* data-pairs are plotted on their respective graphs we obtain the following (figures 3-5 and 3-6):

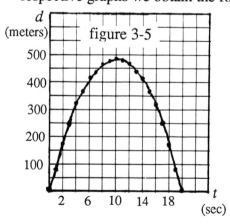
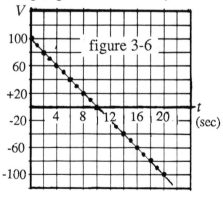

(2)The *d-t* graph below (figure 3-7) represents the data obtained for the motion of a car on a straight highway. During which portion of the graph, if any, was the car: (a) speeding up, (b) at constant speed, (c) slowing down and (d) at rest? (e) What distance did the car travel during the second and third seconds? (f) What was its average velocity during the first three seconds? (g) What was its average velocity during the last three seconds? Assuming the car accelerated and decelerated uniformly whenever its speed changed, what was its speed: (h) at the beginning of the eight second interval depicted in the graph, (i) when the clock read *3 seconds*, (j) at $t = 5$ seconds and (k) at the end of the eight second interval? (l) Draw a *v-t* graph to represent the motion of the car for the entire eight second period of time.

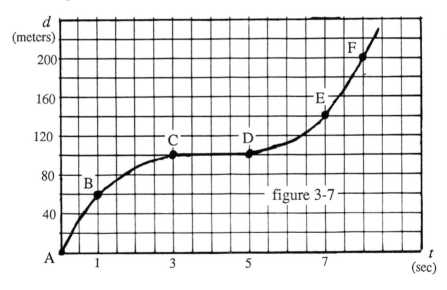

SKETCH AND IDENTIFY

The AC portion of the graph looks very much like graph (g) in section 3.1 - moving away from the reference point, positive velocity, slowing down and a negative acceleration rate. While the particular curvature of such a graph determines whether or not the acceleration rate is uniform and what its value is, and type (g) graphs may have a variety of curvatures, we are told that whenever the speed of the car is changing it occurs at a uniform rate. The value of the acceleration rate for the AC portion of the graph can be found by comparing v_i to v_f for the corresponding time interval (the first three seconds) and using the formula $a = \Delta v/\Delta t$.

The DF portion of the graph is of type (e) - moving away from the reference point, positive velocity, gaining speed, and a positive acceleration rate. Again, we are told that the acceleration rate is uniform during this time interval (the last three seconds).

Since the graph is continuous, with no illegal sharp corners, we know that the final velocity of the AC portion must be the same as the initial velocity of the CD portion, and the final velocity of the CD portion must be the same as the initial velocity of the DF portion. Otherwise the velocity of the car would jump from one value to another without passing through the in-between values. And this jump would take no time at all, since no time elapses between the last point of the AC portion and the first point of the CD portion. Such jumps, called *discontinuities*, are against the rules of classical physics.

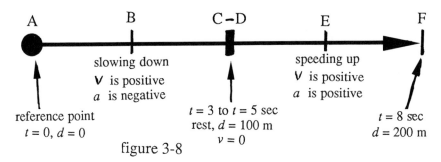

figure 3-8

LINK UNKNOWN TO GIVEN

(a) The car was gaining speed in the DF portion of the graph since it is a type (e) graph. Or, because small *connecting lines* become increasingly vertical, with ever increasing slope, as we go from D to F and the slope on a *d-t* graph, we know, is equal to the average velocity (Formulas, item e).

(b) The answer is CD. It is the only portion of the graph that is straight. All curved *d-t* graphs (e, g, i, k) represent changing speed, and all straight *d-t* graphs (a, c) represent constant speed.

(c) The answer is AC because it is a type (g) graph and the small connecting lines become less steep as we go from A to C.

(d) The answer is CD. The slope is zero and constant, so the velocity must be fixed at zero.

(e) The "second and third seconds" phrase refers to the time interval between $t = 1$ and $t = 3$. During this time the car went from 60 meters to 100 meters from the reference point, in a straight line, in one direction (since the velocity is never negative in the problem). It must therefore have traveled 100 - 60, or 40, meters during that period of time.

(f) $\bar{v} = d_T/t$ (first three seconds)

$\bar{v} = 100/3 = 33.33$ m/s

(g) $\bar{v} = d_T/t$ (last three seconds)

$\bar{v} = (200 - 100)/3 = 33.33$ m/s

(h) $\bar{v} = (v_i + v_f)/2$ (first 3 seconds, uniform acceleration)

$33.33 = (v_i + 0)/2$ ($v_f = 0$, same as v_i for CD)

$v_i = 66.66$ m/s

The velocity at $t = 0$ is 66.66 m/s.

(i) The velocity at $t = 3$ seconds is zero, the same as during the

Motion Graphs

entire CD portion of the graph.

(j) The velocity at $t = 5$ seconds is also zero for the same reason.

(k) $\bar{v} = (v_i + v_f)/2$ (last 3 seconds, uniform acceleration)

$33.33 = (0 + v_f)/2$

$v_f = 66.66$ m/s

The velocity at $t = 8$ seconds is 66.66 m/s.

(l) To prepare a *v-t* graph we must first determine the acceleration rate during the AC and DF portions, otherwise we don't know what slope to draw (slope on a *v-t* graph equals acceleration rate - Formulas, item f).

$v_f = v_i + at$ (AC portion, first three seconds)

$0 = 66.66 + a\,(3)$

$a = -22.22$ m/s^2

$v_f = v_i + at$ (DF portion, last three seconds)

$66.66 = 0 + a\,(3)$ $(t = 8 - 5)$

$a = 22.22$ m/s^2

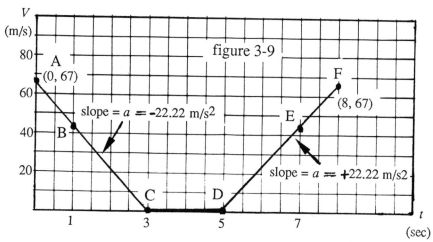

figure 3-9

(3) The velocity-time graph below (figure 3-10) represents the data obtained for the motion of a rocket launched vertically upward. (a) During which portion of the graph is the acceleration rate (in absolute value) the greatest? (b) For how many seconds does the rocket rise? (c) How high does the rocket rise? (d) When does the rocket return to its launching pad, assuming it falls at a uniform acceleration rate? Ignore the effects of air resistance.

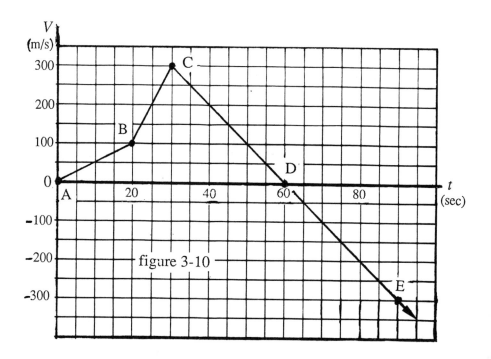

figure 3-10

SKETCH AND IDENTIFY

The AB and BC portions of the graph are of type (f) - positive velocity, moving away from the reference point (which we choose to be the ground), gaining speed and a positive acceleration rate. The only difference between the AB and BC portions is in their slope. The BC portion is obviously steeper than the AB portion.

Motion Graphs 69

The CD portion is an (h) type graph - positive velocity, moving away, slowing down and a negative acceleration rate. Notice that although the graph turns around at point C, the rocket does not - it is still moving away from the ground since its velocity is positive even after point C. This remains the case all the way to point D.

The DE portion is an (l) type graph - negative velocity, moving toward (the ground), gaining speed and a negative acceleration rate. The rocket turns around at point D (when $t = 60$ seconds) since that is the point where the velocity ceases to be positive (moving away) and begins to be negative (moving toward). At the turn-around point, at D, the rocket's velocity is zero.

The sharp corners in the graph, at points B and C, are not to be taken literally. An object's velocity or acceleration rate cannot jump over in-between values, as a literal interpretation of the graph indicates must occur at $t = 20$ and $t = 30$ seconds. The sharp points merely mean that the changeover (in velocity) occured very rapidly, during very small intervals of time, and that the details of the changeover are to be ignored.

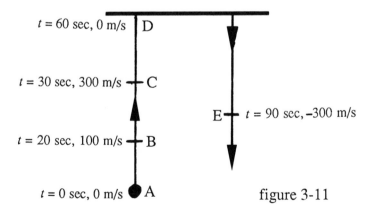

figure 3-11

LINK UNKNOWN TO GIVEN

(a) To find the portion of the graph with the greatest absolute value of acceleration rate, we determine the acceleration rate in each portion and compare.

$a = \Delta v/\Delta t = (100 - 0)/(20 - 0) = +5$ m/s²   (AB)

$a = \Delta v/\Delta t = (300 - 100)/(30 - 20) = +20$ m/s²   (BC)

$a = \Delta v/\Delta t = (-300 - 300)/(90 - 30) = -10$ m/s²   (CD, DE)

The greatest absolute acceleration rate occurs during the BC portion, between $t = 20$ and $t = 30$ seconds.

(b) It rises as long as the velocity is positive (moving away), from $t = 0$ to $t = 60$ seconds.

(c) To find the height achieved by the rocket we must determine the distances traveled during the AB, BC and CD portions and add these together. We can do this either by finding the area under each portion of the graph, or by using the formula $d = v_i t + at^2/2$ (from chapter two) and setting our clock to zero at the beginning of each portion (the symbol d in that formula, you will recall, can also equal the distance traveled between any time t and when $t = 0$, so long as the object travels in one direction). We cannot apply the formula to the first sixty seconds all at once, by plugging $t = 60$ into the formula, because a does not remain fixed throughout the entire sixty second interval.

AREA METHOD

$A_\triangle = bh/2$ (area of triangle = base times height / 2)

$A_{AB} = (20)(100)/2$ (area under AB portion of graph)

$A_{AB} = 1000$ square units

$A_{BC} = A_\square + A_\triangle$ (area of rectangle + triangle on top of it)

$A_\square = bh$

Motion Graphs

$A_{BC} = (10)(100) + (10)(200)/2 = 2000$ square units

$A_{CD} = (30)(300)/2 = 4500$ square units

Total area is $(1000 + 2000 + 4500)$ square units. Total distance traveled upward is 7500 meters.

FORMULA METHOD

$d = v_i t + at^2/2$ (from $t = 0$ to $t = 20$ seconds)

$d_{AB} = 0 + (5)(20)^2/2$

$d_{AB} = 1000$ m

$d_{BC} = (100)(10) + (20)(10)^2/2$ (setting $t = 0$ at point B renders v–initial equal to 100 m/s and $t = 10$ at point C)

$d_{BC} = 2000$ m

$d_{CD} = (300)(30) + (-10)(30)^2/2$ (setting $t = 0$ at point C renders v–initial equal to 300 m/s and $t = 30$ at point D)

$d_{CD} = 4500$ m

Total distance traveled upward is $(1000 + 2000 + 4500)$ or 7500 meters, the same answer we obtained via the area method.

(d) For the rocket to return to its launching point it must fall 7500 meters as it accelerates downward at the rate of 10 m/s² (a convenient approximation of 9.8 m/s²).

If we set our clock to zero at point D (where the downward trip begins) and make the rocket's highest point the reference point for the downward trip, the velocity and acceleration rate going down become positive (the rocket is then moving away from the reference point at ever increasing speed) and d–initial becomes zero. We can then put the distance formula to use as follows:

$d = v_i t + at^2/2$ (downward trip)

$7500 = 0 + (10)t^2/2$

$7500 = 5t^2$

$t = 38.73$ sec

This means that after falling for 38.73 seconds the rocket returns to its starting point. Since these 38.73 seconds begin at point D, when the clock of our graph reads $t = 60$ seconds, the rocket returns when that clock reads 98.73 seconds (beyond point E).

(4) The d-t graph below (figure 3-12) represents the motion of a uniformly accelerating object. Find its acceleration rate.

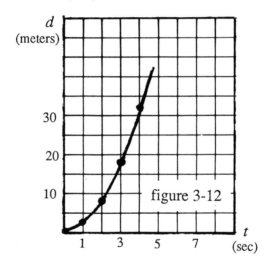

figure 3-12

IDENTIFY AND SKETCH

The acceleration rate is the ratio $\Delta v / \Delta t$. To find this ratio we must have v values for various times, t. Since this data is not provided, we must dig it up. The value of v at any time t can be found by drawing a tangent to the graph at the corresponding point in time and determining its slope. By choosing two such points and comparing their respective v values we can calculate Δv that corresponds to the chosen Δt and hence the ratio $\Delta v / \Delta t$.

Motion Graphs 73

LINK UNKNOWN TO GIVEN

Our arbitrarily chosen points are $t = 1$ and $t = 4$. We draw tangents to the graph at these points, as illustrated in figure 3-13. Care must be taken to draw the tangents properly - each tangent must make contact with the graph at one point only.

The slope of each tangent is determined in the usual manner - by calculating the ratio $\Delta y / \Delta x$. The tangent to the graph at $t = 1$ turns out to have a slope of 8/2, or 4, and the tangent at $t = 4$ has a slope of 16/1, or 16.

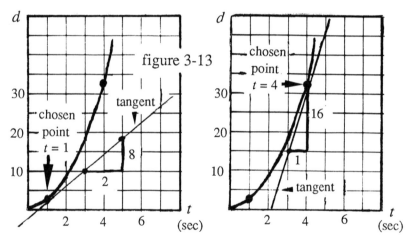

We conclude therefore that the velocity (equal to the slope of the tangent) at $t = 1$ is 4 m/s and the velocity at $t = 4$ is 16 m/s. Thus,

$a = \Delta v / \Delta t$ \qquad (from $t = 1$ to $t = 4$)

$a = (16 - 4)/(4 - 1) = 12/3 = 4$ m/s^2

3.5 ON YOUR OWN PROBLEMS

(Asterisk indicates solution appears in appendices.)

*(1) Find the acceleration rate and distance traveled during the (a) AB, (b) BC and (c) CD portions of the *v-t* graph below (figure 3-14).

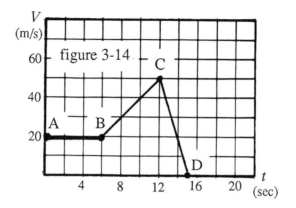

figure 3-14

*(2) A car located 150 meters from an intersection moves away from the intersection at the constant velocity of 30 m/s. After maintaining this velocity for 10 seconds it decelerates to a stop in 5 seconds. It then rests for another 5 seconds, turns around and accelerates toward the intersection at the rate of 2 m/s², until it reaches the intersection.

(a) Draw a *v-t* graph representing the motion described above.

(b) What is the greatest distance between the car and the intersection?

(c) After how much time does the car return to the intersection? Indicate the corresponding point on the graph.

(d) Draw a *d-t* graph representing the motion of the car.

*(3) For each of the three portions of the *d-t* graph below (AB, BC and CD) indicate whether the object (a) is moving toward or away from the reference point, or not at all, (b) has a positive, negative or zero velocity value, (c) has a positive, negative or zero acceleration rate. (d) Assuming each acceleration or deceleration proceeds at a uniform rate, find the acceleration rate for each portion of the graph. (e) Draw a *v-t* graph to represent the motion of the object.

Motion Graphs

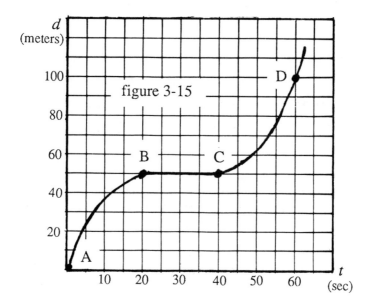

figure 3-15

*(4) Solve *On Your Own* problem five of chapter two by sketching a *d-t* graph for each vehicle on the same set of axes.

(5) Solve *On Your Own* problem twelve of chapter two graphically.

TABLE 3-2

time	distance A	distance B
0	0	0
1	1	2
2	4	8
3	9	18
4	16	32
5	25	50
6	36	72
7	49	98
8	64	128

(time in seconds, distance in meters)

(6) The data above (table 3-2) describe the motion of two cars that started simultaneously from rest and traveled side by side as

they accelerated uniformly for ten seconds. Both cars started from the same point on the road.

Draw the *d-t* graph for each on the same set of axes. Do likewise for their *v-t* graphs. Determine from the graphs: (a) how far apart they are 4.5 seconds after they start moving, (b) the point in time when one of them is traveling 8 m/s faster than the other.

Base your answers to problems 7 thru 12 on the *d-t* graph below (figure 3-16) which represents the motion of an object moving along a straight line.

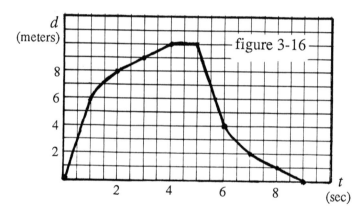

(7) How far did the object travel during the third second?

(8) When and for how long is the object at rest?

(9) During which period of time is the average velocity of the object 3 m/s?

(10) What is the average velocity of the object for the first five seconds?

(11) When and for how long is the object's velocity -1 m/s?

(12) During which time interval(s) is the acceleration rate positive, zero and negative?

- - - - - - -

Motion Graphs

(13) Which of the graphs below (figure 3-17) illustrate each of the following relationships with respect to time: (a) The velocity of a freely falling object, (b) The distance between an observer and a stationary object, (c) The velocity of an object with no acceleration, (d) The distance between the ground and a ball launched upward?

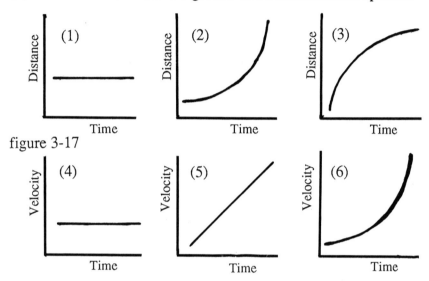

figure 3-17

Problems 14 thru 19 refer to the v-t graph below (figure 3-18) which represents the motion of an object moving along a straight line.

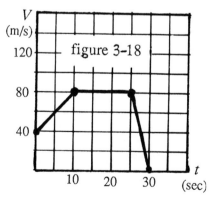

(14) What is the acceleration rate at $t = 5$ seconds? At $t = 20$ seconds? At $t = 28$ seconds?

(15) How far does the object travel during the first ten seconds?

(16) How far does the object travel between $t = 10$ seconds and $t = 25$ seconds?

(17) How far does the object travel in the last five seconds?

(18) If the object's location at $t = 0$ is 200 meters from the reference point, how far from the reference point is the object when the clock reads *30 seconds* ?

(19) Draw a *d-t* graph representing the motion of the object.

- - - - - - -

(20) The slope of a *d-t* graph represents what feature of an object's motion?

(21) The *y*-intercept of a *v-t* graph represents _____.

(22) A negative value for velocity indicates that _____.

(23) Can the *d-t* or *v-t* graph that represents the motion of an object whose direction changed be a straight line? If yes, provide an example by sketching a sample graph.

(24) Can an object be traveling faster and faster yet have a negative acceleration rate? If yes, provide an example.

(25) Sketch a *d-t* and *v-t* graph for an object that is moving toward the reference point and slowing down.

(26) The velocity of an object that is accelerating uniformly is represented by which feature of its *d-t* graph?

- - - - - -

Questions 27 thru 30 refer to the *v-t* graph below (figure 3-19) which represents the motion of two objects, A and B, both of which start moving simultaneously from the same place (at $t = 0$). Both objects travel in the same direction, along a straight line.

Motion Graphs

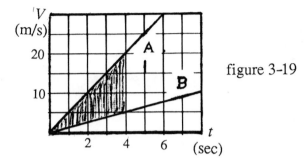

figure 3-19

(27) Which object, A or B, is accelerating at the slower rate?

(28) What distance does object B travel during the first eight seconds?

(29) How long does it take object A to travel the same distance (as object B in question twenty-eight)?

(30) The shaded portion of the graph represents _____.

4

CHAPTER FOUR

VECTORS

4.1 VOCABULARY AND FORMULAS

DEFINITIONS

SCALAR QUANTITY - Quantity that has magnitude but no direction, such as time and mass.

VECTOR QUANTITY - Quantity that has magnitude and direction, such as velocity and force.

VECTOR - A directed arrow used to represent a vector quantity. The direction of the arrow is the same as the direction of the vector quantity and the length of the arrow is proportional to - based on a specified scale - the magnitude of the vector quantity.

EXAMPLE

Vector quantity	**Vector**
Force, 150 N due east	———▶ EAST
Scale: 1 cm = 50 N	3 cm

RESULTANT - The vector sum of two or more vectors.

COMPONENT - Each of two or more vectors whose resultant is

a given vector. Each of these vectors is referred to as a component of the given vector.

SCALE - The relationship between the unit of length used to construct a vector and the amount of vector quantity it represents. For example: 1 cm = 50 N.

VECTOR RESOLUTION - The process of finding a particular set of components for a given vector. A vector may be resolved into many different sets of components.

DISTANCE and DISPLACEMENT - Distance is a scalar quantity. It is the length of the path traversed between two points. Displacement is a vector quantity defined as the arrow directed from the starting point to the ending point.

SPEED and VELOCITY - Speed is a scalar quantity defined as distance traveled per time. Velocity is a vector quantity. Its direction is that of the motion of the object and its magnitude is the speed of the object.

FORCE - A vector quantity loosely defined as a push or pull. The push or pull need not produce motion; it exists if it acts or attempts to change the motion-status of an object, whether or not it succeeds in doing so.

EQUILIBRIUM - Condition in which the resultant of all the forces acting on an object (the *net* force) is equal to zero.

EQUILIBRANT - Force vector which when added to other force vectors present results in a condition of equilibrium. It is equal in magnitude but opposite in direction to the resultant of the vectors present.

CONCURRENT VECTORS - Two or more vectors active at the same time at the same point. Adding vectors is meaningful only if they are concurrent.

FORMULAS

(a) *Head to tail method of vector addition* : Arrange vectors to be added in head to tail fashion. Resultant is vector drawn from the tail of the first to the head of the last (figure 4-1).

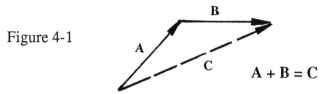

Figure 4-1

$A + B = C$

(b) *Parallelogram method of vector addition* : Arrange vectors to be added (no more than two at a time) in tail to tail fashion. Complete parallelogram. Resultant is diagonal directed from the point where the tails of the vectors meet to the opposite corner, as illustrated (figure 4-2).

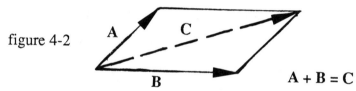

figure 4-2

$A + B = C$

(c) *Vector resolution* : Reverse process of parallelogram method of vector addition. If vector **A** is resolved into two perpendicular components, $\mathbf{A_x}$ and $\mathbf{A_y}$, then $A_x = A \cos \theta$ and $A_y = A \sin \theta$, where A, A_x and A_y (not in bold letters) refer to the magnitude of vectors **A**, $\mathbf{A_x}$ and $\mathbf{A_y}$, respectively (figure 4-3).

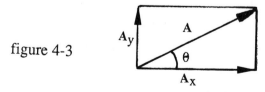

figure 4-3

(d) *Component method of vector addition* : If **C** = **A** + **B**, then $\mathbf{C_x} = \mathbf{A_x} + \mathbf{B_x}$ and $\mathbf{C_y} = \mathbf{A_y} + \mathbf{B_y}$ where the subscripts x and y denote the components of the vector in the x and y directions.

Vectors 83

(e) *Vector subtraction* : If **C** = **A** - **B**, then **C** = **A** + (-**B**) where (-**B**) is a vector with the same magnitude as **B** but in the opposite direction. In other words, if **B** = (**B**$_x$, **B**$_y$) then (-**B**) = (-**B**$_x$, -**B**$_y$).

4.2 THINGS TO KNOW

(1) Vectors are not numbers and cannot be added or subtracted as such. Think of a vector as a picture. When one picture is added to another, the result is a third picture - certainly not a number. The same is true of vectors. When two or more vectors are added the result is not a number but another vector called the *resultant.* To find the resultant the rules of vector addition must be used (Formulas, items *a, b* and *d*).

An important consequence of this is that vector equations such as **A** + **B** = **C** cannot be solved algebraically. You cannot claim, for example, after determining that vector **A** is 40 N strong and vector **B** is 20 N strong, that vector **C** is 40 + 20, or 60 N, strong. It doesn't work that way - vectors **A** and **B** may not be pointing in the same direction. You must resort to the rules of vector addition.

To summarize: When you see the vector symbol, such as **A**, think picture, not number! (Vector quantities in this book will be designated as such by **bold** letters. Whenever a vector quantity does not appear in bold, such as A, the reference is to *the magnitude of* the vector quantity, which is a scalar.)

(2) Vectors drawn to represent specific vector quantities (such as a particular velocity or force) must be drawn *to scale.* That is, a specific scale must be chosen delineating the relationship between the unit of length on the vector and a particular amount of vector

quantity. This must be done *before* the vector is drawn because the length of the vector is based on it. For example, to represent a 20 N force we might choose a scale of 1 cm = 4 N, then draw a 5 cm long arrow to represent the force. If two or more vectors are to be added or subtracted, they must all be drawn based on the same scale.

(3) When the resultant of two or more vectors is found by vector addition, remember to reconvert the length of the resultant to an amount of vector quantity. This conversion must, of course, be based on the scale used to draw the vectors that were added together.

(4) Forces may be added only if they are concurrent, that is, they act at the same point, at the same time.

(5) The effect of two or more concurrent forces acting on an object is the same as if their resultant acted on the object.

4.3 SOLVING THE PROBLEMS

(1) The engine of an airplane propels it due east at a velocity of 100 m/s, at the same time that a wind acts to push the airplane due south at the rate of 50 m/s. What is the airplane's actual velocity (speed and direction) as seen from the ground?

SKETCH AND IDENTIFY

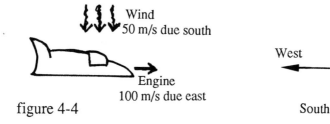

figure 4-4

Vectors

The two vector quantities in this problem may be represented by vectors, as such:

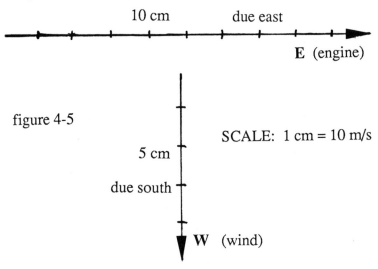

figure 4-5

SCALE: 1 cm = 10 m/s

LINK UNKNOWN TO GIVEN

To find the resultant of these vectors we employ the head to tail method and solve two ways - first by construction (also called graphing), then by calculation.

BY CONSTRUCTION

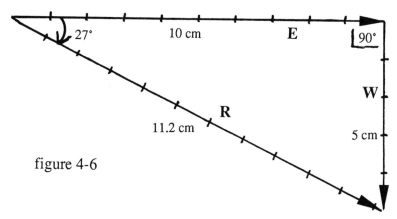

figure 4-6

Since 11.2 cm represents 112 m/s based on our chosen scale,

the resultant velocity of the airplane is 112 m/s directed 27° south of east. This should be the airplane's actual speed and direction, as seen from the ground.

When adding vectors by construction it is imperative that all vectors be drawn and oriented properly. If any vector is made too long or too short, or if the angle between them is not drawn correctly, the magnitude and direction of the resultant obtained may turn out to be wrong. The more accurately the vectors are drawn, the more accurate will be the resultant's length and direction. In the above example, vector **E** must be drawn 10 cm long, vector **W** must be 5 cm long and the angle between **E** and **W** must be 90°. The resultant's length and direction are obtained by measurement, using a ruler and a protractor.

BY CALCULATION

We can spare ourselves the task of constructing vectors to scale and measuring lengths and angles if we're willing to apply our imagination and do some calculating. We simply *make believe* that vector **E** below (figure 4-7) is 100 cm long, with each cm representing 1 m/s. We also *make believe* that vector **W** (same figure) is 50 cm long and that the angle between the vectors is 90°. Then we draw **R** and calculate (not measure!) what its length, x, and direction, θ, are supposed to be, as follows:

$x^2 = 100^2 + 50^2$ (pythagorean theorem)
$x^2 = 12{,}500$
$x = \sqrt{12{,}500} = 111.8$ cm
$x = 111.8$ m/s
$\tan \theta = 50/100 = .5$
$\theta = 27°$

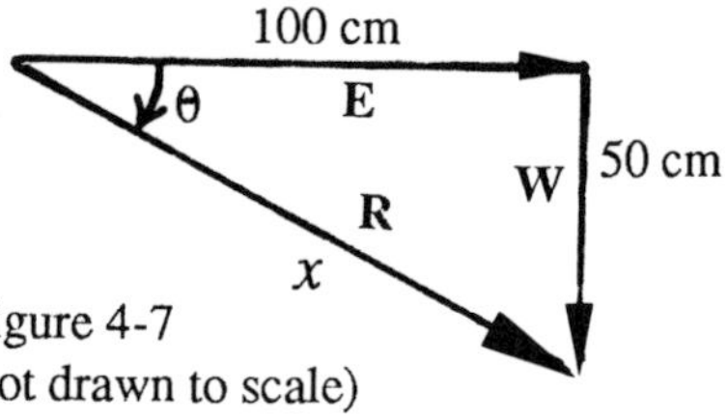

figure 4-7
(not drawn to scale)

Vectors

Determining the resultant by calculation offers a distinct advantage - it provides more precise and more dependable solutions. In some cases, however, the mathematics involved can become quite complicated, such as when adding three or more vectors simultaneously. You should know how to add vectors both ways - by construction and by calculation.

It should be noted that the order in which the vectors are arranged is irrelevant. It does not matter whether we first draw vecter **E** and then place the tail of **W** at the head of **E** (case *a*, figure 4-8) or first draw vector **W** and then place the tail of **E** at the head of **W** (case *b*, same figure). The resultant in either case is the same - a

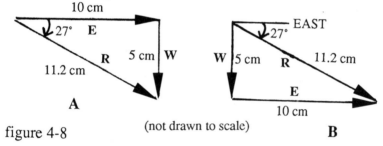

figure 4-8   (not drawn to scale)

vector 11.2 cm long, directed 27° south of east. This is always the case, irrespective of how many vectors are added, so long as each vector is drawn correctly - with the correct length and direction - and the tail of each successive vector to be drawn is placed at the head of the previous vector.

(2) Two people push simultaneously on the same rock, each with a force of 20 Newtons. What is the magnitude and direction of the resultant force if the angle between the forces is (a) 30°, (b) 135°, (c) 0° and (d) 180° ? (e) When the angle between the forces is 30°, how hard and in what direction must a third person push in order that the rock be in a state of equilibrium?

SKETCH AND IDENTIFY

Since the only angle identified in the problem is that between the forces and no mention is made of a particular direction for either force, we may choose any direction for the forces so long as the angle requirement is met. Let us choose the direction of one 20 N force to be due east and label this force vector quantity **A**. On a scale of 1 cm = 4 N this vector quantity is represented by vector **A** as such:

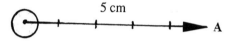

The other vector quantity, **B**, on the same scale, is then represented as illustrated below:

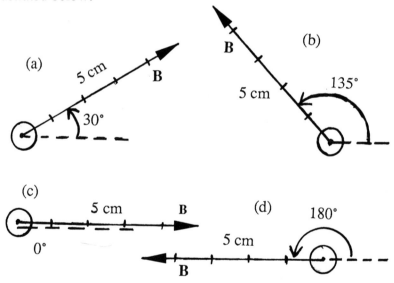

LINK UNKNOWN TO GIVEN

Since the forces act concurrently they may be added, using the rules of vector addition, to find their resultant. In this problem we employ the parallelogram method.

Vectors

BY CONSTRUCTION

(a) 30°   1 cm = 4 N

figure 4-9

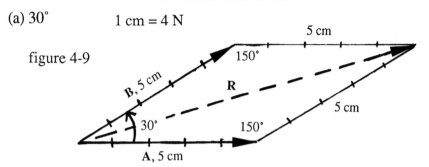

Step One: Arrange vectors tail to tail. Make sure their lengths and orientations are as accurate as possible.

Step Two: Complete the parallelogram. Make sure opposite sides are truly parallel. The best way to do this is to construct supplementary consecutive angles. (The lower right angle should be 150°, as is the case with the upper left angle. 150+30 = 180.)

Step Three: The resultant is the diagonal directed from the meeting point of tails to the opposite corner. Obtain the direction and magnitude of the resultant by measurement. In figure 4-9 the resultant turns out to be 9.65 cm long, representing - based on the scale employed - a 38.6 N force, directed 15° north of east. (Check it out!)

BY CALCULATION

Step One: Sketch vectors tail to tail making believe each is 20 cm long (based on a scale of 1 cm = 1 N) and that the angle between them is 30°.

(not drawn to scale)
figure 4-10

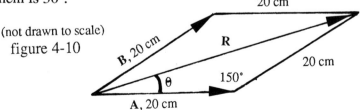

Step Two: Sketch parallelogram and diagonal, as in figure 4-10.

Step Three: Calculate the length and direction of the diagonal. This diagonal is the resultant, directed from the meeting point of tails. Since the two triangles enclosed in figure 4-10 are not right triangles, the pythagorean theorem cannot be used. Nor can any of the six trigonometric functions (sine, cosine, etc.) be used. Instead, we turn to the *Law of Cosines*.

To find Magnitude of R:

$$R^2 = 20^2 + 20^2 - (2)(20)(20)(\cos 150°)$$
$$R^2 = 400 + 400 - (800)(-.866)$$
$$R = 38.6 \text{ cm} = 38.6 \text{ N}$$

To find direction of R:

$$20^2 = 38.6^2 + 20^2 - (2)(38.6)(20)(\cos \theta)$$
$$-38.6^2 = -(2)(38.6)(20)(\cos \theta)$$
$$38.6 = 40 \cos \theta$$
$$.965 = \cos \theta$$
$$\theta = 15°$$

(b) 135°

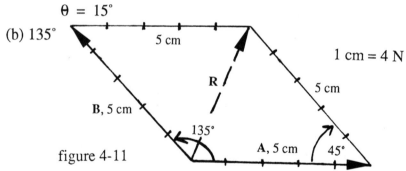

figure 4-11

R is 3.83 cm long and 15.32 N strong, directed 67° north of east.

(c) 0°. The parallelogram method cannot be used when the angle is 0° or 180° or when three or more vectors are to be added simultaneously. Using the head to tail method we proceed as follows:

Vectors

1 cm = 4 N

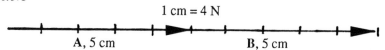

A, 5 cm   B, 5 cm

The resultant is a vector drawn from the tail of the first to the head of the last, as such:

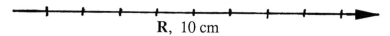

R, 10 cm

Thus the resultant is 40 N strong, directed due east.

(d) 180°. Again, the head to tail method must be used, as such:

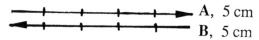

A, 5 cm
B, 5 cm

1 cm = 4 N

Since **A** and **B** are supposed to overlap, a vector drawn from the tail of **A** to the head of **B** has no length at all, and the resultant is zero Newtons strong.

(e) 30°. To create a state of equilibrium we introduce an equilibrant vector, one that is equal in magnitude but opposite in direction to the resultant of the vectors present. Since the resultant of **A** and **B** when they are 30° apart is 38.6 N strong, directed 15° north of east (see solution to part *a* above), we proceed as illustrated in figure 4-12.

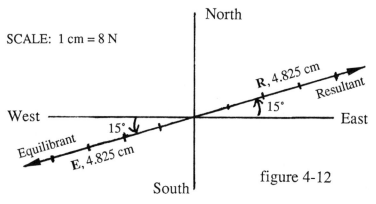

figure 4-12

The equilibrant, **E**, is 4.825 cm long and - based on the scale employed - 38.6 N strong, directed 15° south of west.

To verify that vector **E** (38.6 N strong, directed 15° S of W) is indeed an equilibrant to vectors **A** and **B** when the angle between them is 30°, we add all three vectors, **A**, **B** and **E**, together to see if the resultant of all three of them is zero (the requisite condition for equilibrium). To do this we must use the head to tail method, as illustrated in figure 4-13.

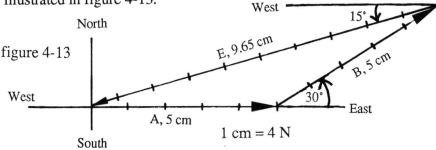

figure 4-13

The resultant of all three vectors is a vector drawn from the tail of the first, **A**, to the head of the last, **E**. Such a vector is zero units long, since the head of **E** is at the tail of **A**.

Figure 4-13 suggests a general conclusion about forces in equilibrium: their vectors form a closed figure when drawn head to tail and to scale.

(3) A 1000 meter wide river flows due south. A woman swims due west, aiming to get across the river. If her swimming speed in still water is 5 m/s and the speed of the current is 3 m/s (that is, the current carries objects in its path downstream at the rate of 3 m/s), (a) how far downstream does she arrive on the other side of the river? (b) How long does it take her to reach the other side? (c) In what direction should she swim in order to arrive at the point directly opposite her starting point?

Vectors

SKETCH AND IDENTIFY

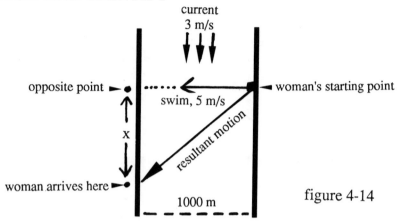

figure 4-14

There are two concurrent velocity vectors in this problem, one due to the woman's swimming effort and the other imposed on her by the current. The former we label **W**. It is directed due west and its magnitude is 5 m/s. The latter we label **C**. It is directed due south and its magnitude is 3 m/s. On a scale of 1 cm = 1 m/s the two vectors representing these vector quantities are:

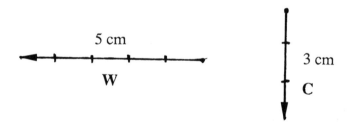

The actual motion of the woman across the river, as seen by someone standing on either of the river's banks, is the resultant of these two vectors. Let us call this resultant **V**. Once we determine the direction and magnitude of **V** we should be able to solve parts *a* and *b* with the help of a little arithmetic and the formula $d = vt$ (from chapter two).

LINK UNKNOWN TO GIVEN

(a)
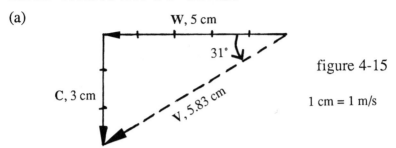

figure 4-15

1 cm = 1 m/s

The woman's resultant velocity is 5.83 m/s, directed 31° south of west.

Now we superimpose this information on the spatial arrangement within the river (figure 4-16).

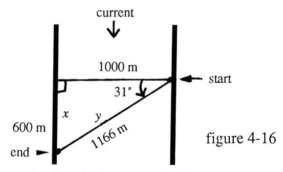

figure 4-16

The distance x can be found trigonometrically:

Tan 31° = $x/1000$ = .6 x = 600 m

The woman arrives 600 meters downstream.

(b) The woman swims y meters (figure 4-16) as she crosses the river in the direction of 31° S of W, at the rate of 5.83 m/s. To find distance y we proceed as follows:

Cos 31° = $1000/y$ = .857 y = 1166 m

To find the time it takes to cross the river:

$d = vt$
$1166 = 5.83t$
$t = 200$ sec

Notice that this is the same amount of time it would take to cross the river had there been no current. Then she would be moving directly across the river, a distance of 1000 meters, at the rate of 5 m/s. The time to cross the river would then satisfy the relationship:

$$d = vt$$
$$1000 = 5t$$
$$t = 200 \text{ sec}$$

In other words, the current neither helps nor hinders her effort to cross the river; it neither reduces the time to get across, nor increases it. This is due to the fact that the current is directed perpendicularly to the direction of swim - no component of the current exists parallel to the direction of the woman's swimming effort. The current merely acts to move her southward at the same time that her swimming effort acts to move her westward.

(c) Let us label the unknown swimming vector we are looking for, **X**. When **X** is added to the current vector **C** we want the result to be that the woman arrives on the other side of the river at the point that is directly opposite her starting point. We therefore insist that

$$\mathbf{X} + \mathbf{C} = \mathbf{V}$$

where the magnitude of **X** is 5 m/s (the woman's swimming speed) and its direction is unknown. Vector **C**, as before, has a magnitude of 3 m/s, directed due south. Vector **V** (the resultant) is of unknown magnitude but must be directed due west (perpendicular to the current) in order that the woman arrive where we want her to - at the point directly opposite her starting point. Solving for **X**, we proceed as follows:

$$\mathbf{X} = \mathbf{V} - \mathbf{C}$$
$$\mathbf{X} = \mathbf{V} + (-\mathbf{C})$$

where vector (-**C**) has the same magnitude as **C** but points in the opposite direction (Formulas, item *e*). Vector (**-C**), therefore, has a magnitude of 3 m/s, directed due north.

BY CONSTRUCTION

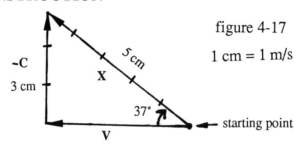

figure 4-17

1 cm = 1 m/s

As can be ascertained from figure 4-17, if the woman were to aim her swimming effort of 5 m/s in the direction of 37° north of west, while the current continued to act to move her southward at the rate of 3 m/s, she would be seen moving straight across, perpendicularly to the current, and will arrive at the point that is directly opposite her starting point. This can be verified by adding **X** to **C**. The resultant is **V**, directed due west.

(4) A 200 N crate is held on a frictionless incline by a rope parallel to the incline, as illustrated (figure 4-18). The angle of inclination is 30°. How strong is the force exerted by the rope on the crate (the tension in the rope)? How strong is the force exerted by the incline against the crate (the normal force)?

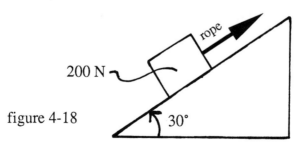

figure 4-18

Vectors 97

SKETCH AND IDENTIFY

The problem says "the crate is held", meaning it is not moving. This implies that the various forces acting on the crate produce a state of equilibrium. There are three forces acting on the crate. One is the pull of gravity (also called weight). It is 200 Newtons strong and acts vertically downward. Another is exerted by the rope as it acts to pull the crate parallel to the incline, toward the top of the incline. This force cannot be equal to the weight of the crate since it does not act opposite to that weight (vertically upward). In other words, the pull of the rope and the pull of gravity cannot possibly create equilibrium in this problem by themselves. There must be a third force. That force is exerted by the incline as it acts to push the crate perpendicularly to the surface of the incline. The three forces are illustrated in figure 4-19.

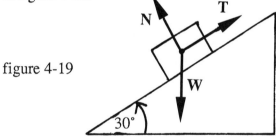

figure 4-19

Now, the tension and normal forces are *response* forces. They respond to other forces by acting to oppose them to the extent necessary to negate ("cancel out") those other forces. Since the force of gravity cannot succeed in its attempt to pull the crate vertically downward (the incline is in the way), it tries to get the crate to slide down the incline. This can only happen if the rope (held fixed at the other end) is stretched and the rope resists being stretched. The rope fights back by acting to pull the crate parallel to and toward the top of the incline. This resistance to being stretched is known as tension. It is only as strong as it needs to be to satisfy its goal - preventing the crate from sliding down the incline. (There is, of course, a limit to the tension any particular rope can exert. If, to satisy its goal of not being stretched, the tension must exceed this maximum, the rope snaps.)

Since gravity cannot get the crate to slide down the incline (the rope opposes that) it tries to move the crate in a manner that does not involve stretching the rope, yet brings the crate closer to the ground. Pulling the crate perpendicularly to the rope (and the surface of the incline) meets no oppposition from the rope. But it does meet opposition from the incline. The incline is made of some rigid material that resists being compressed, deformed, bent, twisted or broken. The incline's rigidity fights back by pushing the crate in the opposite direction - perpendicularly to and away from the surface of the incline. This resistance to compression is referred to as the *normal* (meaning perpendicular) force. It too is only as strong as it needs to be to accomplish its goal - preventing the crate from moving perpendicularly to and into the incline. (This force also has a maximum, based on the composition of the incline. Exceed the maximum and the incline gives way and collapses.)

To find the strength of each of these response forces in our problem, let us look at exactly what they are responding to. Assume there is no rope and the crate is free to slide down the incline. How strong is the force acting to pull the crate down the incline? It is not 200 N - that's the pull of gravity *vertically* downward. Only a component of gravity's force acts parallel to the incline, toward the bottom of the incline. Analogously, assume the incline suddenly turns soft and the crate is free to move into and through it (perpendicularly to the rope which has not lost its strength). How

Vectors 99

strong is the force in the direction perpendicular to the rope? Again, it is not 200 N, only a component of it.

What we need to do is resolve the vertically acting 200 N vector, **W**, into two perpendicular components, as sketched in figure 4-20. Let us label one component $W_{\parallel}$ (parallel to the incline) and the other $W_{\perp}$ (perpendicular to the incline). The tension **T** responds to and therefore is equal in magnitude and opposite in direction to $W_{\parallel}$, while the normal **N** responds to and is therefore equal and opposite to $W_{\perp}$.

Our scenario for equilibrium is now complete. Vector **W** consists of and can be replaced by $W_{\parallel}$ and $W_{\perp}$. Then **T** and $W_{\parallel}$ are made to add up to zero (equal and opposite) and **N** and $W_{\perp}$ are made to add up to zero (also equal and opposite). Zero plus zero adds up to an even bigger zero, the sum of all the vectors is zero, and we have equilibrium.

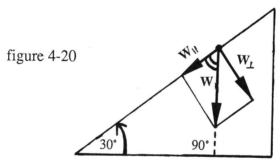

figure 4-20

LINK UNKNOWN TO GIVEN

To resolve **W** into two perpendicular components we use the parallelogram method of vector addition in reverse. In other words, we seek two perpendicular vectors which, when arranged tail to tail, will form a parallelogram with the vector we are resolving as diagonal. In this problem we have an additional condition: The

100 How To Study Physics

orientation of the two perpendicular vectors must be based on the incline - one parallel to the incline, the other perpendicular to it.

We proceed in steps, by calculation:

Step One: Draw vector **W** to become the diagonal.

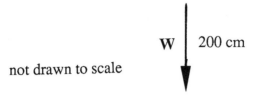

not drawn to scale

Step Two: Draw the desired perpendicular directions from the tail of **W** (figure 4-21). The angle between **W** and the parallel-to-the-incline direction is 60°, and the angle between **W** and the perpendicular-to-the-incline direction is 30°. This must be so since

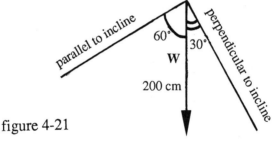

figure 4-21

the angle between **W** and the base of the incline is 90°, the angle of inclination is 30°, and the angles of a triangle must add up to 180° (figure 4-22).

figure 4-22

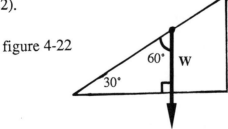

Step Three: Complete the parallelogram by drawing lines from the head of **W** parallel to each of the two perpendicular directions (figure 4-23).

Vectors

figure 4-23

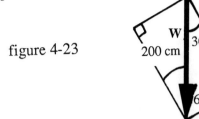

Step Four: The two sides of the parallelogram that meet at the tail of **W** are now turned into vectors $W_\parallel$ and $W_\perp$, the two perpendicular components of vector **W**. The tails of these vectors should meet at the tail of **W** (figure 4-24).

figure 4-24

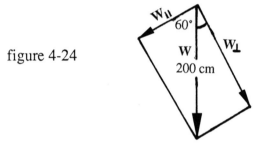

Step Five: To make sure we're on the right track, verify that if $W_\parallel$ and $W_\perp$ are added together, using the parallelogram method of vector addition, the resultant is the given vector **W**.

We now calculate the magnitudes of $W_\parallel$ and $W_\perp$ (we already know their directions) as follows:

$$\cos 60° = W_\parallel/200 = .5 \qquad \cos 30° = W_\perp/200 = .866$$
$$W_\parallel = 100 \text{ N} \qquad\qquad W_\perp = 173 \text{ N}$$

The tension **T** is therefore 100 N strong, directed parallel to the incline, toward the top of the incline, and the normal force **N** is 173 N strong, directed perpendicularly to and away from the incline.

We can check our answers by drawing the three vectors that act on the crate, **W**, **N** and **T**, to scale and arranging them head to tail. If the three of them together create a state of equilibrium, as we claim they do, the arrangement should form a closed figure (see problem two). That they indeed do so is evident from figure 4-25.

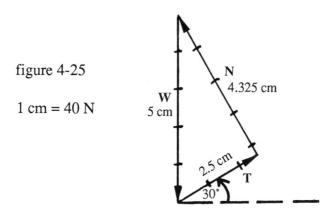

figure 4-25

1 cm = 40 N

(5) A picture weighing 1000 N hangs on a wall. It is supported by two wires each of which makes an angle of 60° with the vertical, as illustrated (figure 4-26). What is the tension in each wire? Would the tension be greater, smaller, or remain the same if the angle with the vertical were changed to 30°?

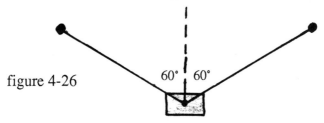

figure 4-26

SKETCH AND IDENTIFY

There are three forces acting on the picture: the vertical-downward pull of gravity of 1000 N and the two forces of tension. Let us label these forces **W**, **T**$_1$ and **T**$_2$, respectively. The three forces must create a state of equilibrium since the picture "hangs" on the wall. It is neither falling, rising, or swaying sideways. The three forces can be represented by vectors as illustrated (figure 4-27).

Vectors

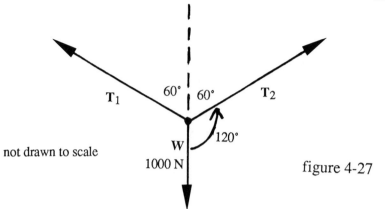

figure 4-27

Since equilibrium exists, we may claim that $\mathbf{W} + \mathbf{T}_1 + \mathbf{T}_2 = \mathbf{0}$. Although this is a vector equation and cannot be solved algebraically, we have the right to rearrange the terms in the following manner: $\mathbf{T}_1 + \mathbf{T}_2 = -\mathbf{W}$. This means that the resultant of $\mathbf{T}_1$ and $\mathbf{T}_2$ must be a 1000 N vector directed vertically upward - very useful information, indeed.

LINK UNKNOWN TO GIVEN

Step One: Let us draw the directions of $\mathbf{T}_1$ and $\mathbf{T}_2$ (not knowing how long these vectors are supposed to be), as in figure 4-28.

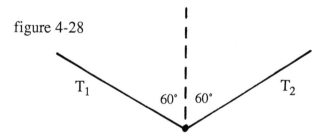

figure 4-28

Step Two: Now we draw their resultant $\mathbf{R}$, based on our knowledge of its magnitude and direction - it is 1000 N strong, directed vertically upward (figure 4-29).

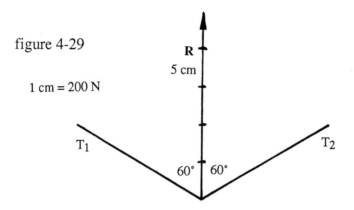

figure 4-29

1 cm = 200 N

Step Three: Next we turn **R** into the diagonal of a parallelogram with T_1 and T_2 as adjacent sides. This is accomplished by drawing lines from the head of **R** parallel to each tension line (figure 4-30).

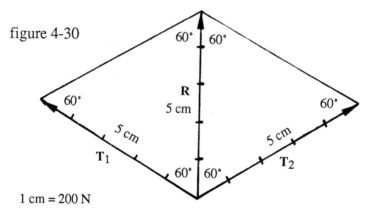

figure 4-30

1 cm = 200 N

Step Four: Now T_1 and T_2 have definite, known lengths. Each turns out to be 5 cm long. Based on our chosen scale this means each represents a tension force of 1000 N.

Why, you may wonder, are *two* 1000 N forces necessary, in this case, to hold up *one* 1000 N weight? Because the forces aren't pulling vertically upward. Had they done so, each would need be only 500 N strong. Instead, the wires are wasting force by pulling sideways, one to the right, the other to the left. They are also

Vectors

fighting each other, since their horizontal efforts are oppositely directed.

The more vertically oriented the wires are, the less force is wasted in unnecessary and contradictory horizontal pulls, so the tension needed to hold up the weight decreases - down to 500 N each when both wires are perfectly vertical. We therefore predict that if the angle between each wire and the vertical is changed to 30°, the tension in each wire would be less than in the case of 60°. This conclusion is verified in figure 4-31.

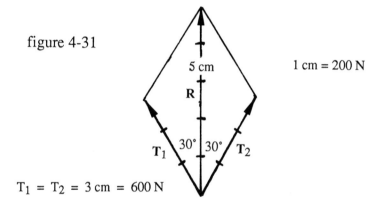

figure 4-31

5 cm

R

1 cm = 200 N

T_1 30° | 30° T_2

$T_1 = T_2 = 3$ cm $= 600$ N

The above described procedure can be applied even when the angles with the vertical are not equal to each other. The problem can also be solved by calculation, rather than construction.

ALTERNATE METHOD

SKETCH AND IDENTIFY

When an object is in a state of equilibrium not only do all the forces acting on it add up (as vectors) to zero, it is also true that all the components in the x direction add up to zero and all the components in the y direction add up to zero. Indeed, if the latter statement is not true, the former cannot be true. This provides us with another method for solving the above problem. We resolve each of the three vectors, **W**, **T**₁ and **T**₂, into x and y components (x and y being perpendicular directions) and set up equations based on the fact that all the x components add up to zero and all the y components add up to zero.

First we choose symbols for all the components: W_x, W_y, T_{1x}, T_{1y}, T_{2x} and T_{2y}. We know that $W_x = 0$ and that $W_y = 1000$. The components of **T**₁ and **T**₂ obey the following relationships, obtained by vector resolution:

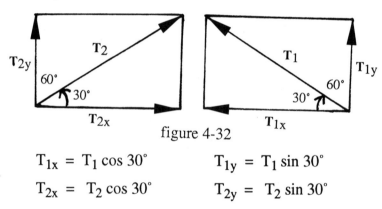

figure 4-32

$T_{1x} = T_1 \cos 30°$ $T_{1y} = T_1 \sin 30°$

$T_{2x} = T_2 \cos 30°$ $T_{2y} = T_2 \sin 30°$

LINK UNKNOWN TO GIVEN

Since all the x components add up to zero we may write:

$$W_x + T_{2x} + T_{1x} = 0$$

Since all x component vectors are directed in the same or in opposite directions, the magnitude of their resultant (the left side of the above equation) is equal to the sum of the magnitudes of the components directed one way, minus the magnitudes of the components directed the opposite way. In this case, we subtract those directed leftward from those directed rightward.

Vectors

$$0 + T_2 \cos 30° - T_1 \cos 30° = 0$$
$$T_2 \cos 30° = T_1 \cos 30°$$
$$T_2 = T_1$$

Since all the *y* components add up to zero:

$$\mathbf{W_y + T_{1y} + T_{2y} = 0}$$

Here we subtract the magnitudes of the *y* components directed downward from those directed upward.

$$-1000 + T_1 \sin 30° + T_2 \sin 30° = 0$$
$$T_1 \sin 30° + T_2 \sin 30° = 1000$$
$$(T_1 + T_2) \sin 30° = 1000$$
$$T_1 + T_2 = 2000$$

Since $T_1 = T_2$ we may dispense with the subscripts and refer to each of them simply as T:

$$T + T = 2000 \qquad 2T = 2000 \qquad T = 1000$$

Thus we conclude that $T_1 = 1000$ N and $T_2 = 1000$ N.

This procedure can also be used when the angles between each wire and the vertical are not equal to each other.

(6) What is the largest weight that can be supported by the structure illustrated in figure 4-33 if the maximum tension the rope can withstand is 2000 N (it snaps if the tension exceeds this amount) and the maximum compression the strut can withstand is 3000 N ?

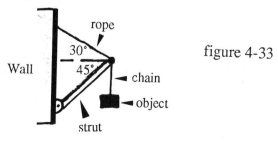

figure 4-33

108   How To Study Physics

Neglect the weight of the rope and strut and assume the vertical chain holding the weight can support any load.

SKETCH AND IDENTIFY

There are three forces active in this problem, and together they must lead to a state of equilibrium. There is the force of gravity, **W**, acting vertically downward on the hanging object. In its attempt to pull the weight down gravity acts to stretch the rope and compress the strut. The rope responds by fighting back with a force of tension, **T**, directed as illustrated in figure 4-34. The strut responds by fighting back with a force of compression, **C**, directed as indicated in the same figure. (For a more complete discussion of tension and compression see problem four.) These forces act parallel to the length of the rope and strut, respectively. No rope can pull perpendicularly to its length, nor does the strut in this problem exert a force perpendicularly to its length (it cannot oppose an attempt to rotate it around the pivot point, where it meets the wall). These forces are concurrent since they converge at the point where the rope, strut and chain meet. The equation **W + T + C = 0** must, therefore, hold true.

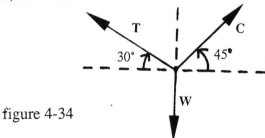

figure 4-34

LINK UNKNOWN TO GIVEN

It is tempting to jump to the conclusion that the largest supportable weight is that supported by the rope and strut when their

Vectors   109

forces of tension and compression are at their maxima - 2000 N and 3000 N, respectively. Acting on this assumption we would proceed to find the magnitude of **W** that leads to equilibrium when the magnitude of **T** is 2000 N and the magnitude of **C** is 3000 N. If we use the *head to tail* method and draw a vector **W** that produces a closed figure (a necessary condition for equilibrium - see problem two) under these conditions, we obtain the vector diagram of figure 4-35.

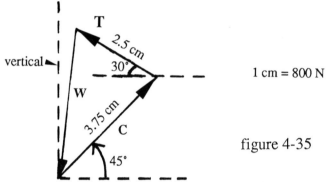

figure 4-35

Since **W** must be directed vertically downward and the **W** vector in the figure is not so directed, the diagram represents an impossible configuration. It is not possible for equilibrium to exist with the tension being equal to 2000 N simultaneous with the compression force being equal to 3000 N, directed as these forces are, no matter what weight we place on the chain. We must therefore seek **T** and **C** vectors with magnitudes below their allowed maxima - yet the weight they support be the maximum possible. Vectors **T** and **C** thus become unknown quantities; we don't know what magnitudes they must have to meet all the conditions of the problem.

Problems such as this are quite difficult to solve by construction. So we turn to mathematical techniques. Let's look at the components. At equilibrium it must be true that:

$C_x + T_x + W_x = 0$ (equilibrium in the x - horizontal - direction)

$C_y + T_y + W_y = 0$ (equilibrium in the y - vertical - direction)

Since all x component vectors are directed in the same or in opposite directions, the magnitude of their resultant (the left side of the first equation) is equal to the sum of the magnitudes of the components directed one way (rightward), minus the magnitudes of the components directed the opposite way (leftward). This leads to the equation:

$C \cos 45° - T \cos 30° + 0 = 0$ (see figure 4-34)

In the y direction (second equation), we add the magnitudes of the components directed upward, then subtract the magnitudes of the components directed downward. This yields:

$C \sin 45° + T \sin 30° - W = 0$

Next we plug in trigonometric values and manipulate the terms in the above equations and obtain:

$.71 \, C = .87 \, T$

$W = .71 \, C + .5 \, T$

We see again that making C equal to 3000 N and T equal to 2000 N is not a possible solution. Such a combination is disallowed by the equation $.71 \, C = .87 \, T$. Instead, let us solve for C in terms of T:

$C = (.87/.71) \, T$

$C = 1.23 \, T$

Plugging this into the above equation for W yields:

$W = .71 \, (1.23 \, T) + .5 \, T$

$W = .87 \, T + .5 \, T$

$W = 1.37 \, T$

Since T cannot exceed 2000 N, W cannot exceed 1.37 times 2000, or 2740 N. When T is at its allowed maximum of 2000 N, W has a magnitude of 2740 N and C is equal to 2500 N - a number that is

Vectors

within its allowed range. These numbers therefore constitute the solution to our problem. The maximum weight that can be supported by the rope and strut, in a state of equilibrium, is 2740 N. When this weight is placed on the chain, the force of tension is 2000 N strong - its maximum - and the force of compression is 2500 N strong. (If you are wondering why we didn't express W in terms of C and then set C at its maximum of 3000 N, well, try it! T would then have to be 2440 N - a number that exceeds its allowed maximum. This is therefore not an allowed combination.)

4.4 ON YOUR OWN PROBLEMS

(Asterisk indicates solution appears in appendices.)

*(1) A hiker walks 10 km due north, then 20 km due west, then 10 km directed 30° south of west. What is her resultant displacement from the starting point?

*(2) Two soccer players kick the ball at exactly the same time. One player's foot exerts a force of 50 N due north, while the other's foot exerts a force of 100 N due east. What is the magnitude and direction of (a) the resultant force on the ball, (b) the equilibrant of the two forces ?

*(3) When a train moves at the rate of 40 km/hr eastward, raindrops that are falling vertically with respect to the earth make traces on the windows of the train that are inclined 30° from the vertical. What is the velocity of the falling raindrops with respect to the earth ?

*(4) An airplane pilot wishes to fly due north. A wind of 50 km/hr is blowing due east. If the flying speed of the airplane (speed in still air) is 120 km/hr, in what direction should the pilot head ?

*(5) A box is dragged across a wooden floor with a rope that forms a 60° angle with the floor. The rope is pulled with a force of 200 N. How much of this force acts to pull the box along the floor?

*(6) How strong must be the force of friction between a 10 N book and a wooden plank to prevent the book from sliding down the plank if the plank is inclined at an angle of 50° from the horizontal?

*(7) A traffic sign is held by two wires connected to two poles. One wire forms an angle of 30° with the horizontal, the other forms an angle of 60° with the horizontal. If the sign weighs 1000 N, what is the tension in each wire?

*(8) What is the maximum weight that can be supported by the two wires in problem seven if the maximum tension each can withstand is 4000 N?

(9) A boy walks 40 meters due south, then 100 meters due west, then 300 meters due north. (a) What is the magnitude and direction of his displacement from the starting point? (b) If the order of the above trips is reversed, what is the displacement from the starting point?

(10) To drag a loaded wagon on a sidewalk a horizontal force of 200 N is required. Two boys pull on horizontal ropes tied to the wagon, each exerting the same amount of force as the other. The ropes are separated by 90°. How much force must each boy exert? Solve by construction.

(11) A sled rope is 5 meters long. A girl holds one end of the rope at a height of 3 meters above the level of the sled and pulls with a force of 40 N. How much force is acting to move the sled horizontally along the floor?

(12) A man weighing 300 N sits at the midpoint of a tight rope that is supported by two poles situated 20 meters apart. The rope

sags 4 meters under the man's weight. What is the tension in the rope?

(13) A woman sits in a hammock whose ropes make 45° angles with the vertical trees supporting either end. The tension in the rope is determined to be 300 N. How much does the woman weigh?

(14) A 500 meter wide river flows north to south. A man swims directly west but arrives 200 meters downstream, on the opposite bank of the river. The man's swimming speed in still water is .5 m/s. (a) How long does it take him to cross the river? (b) With what speed is the current carrying him downstream? (c) What is his actual speed across the river? (d) In what direction must he swim in order to arrive at the point directly opposite his starting point?

(15) An airplane flies at the rate of 400 km/hr, directed 60° south of east. How fast is the airplane traveling due east from its starting point?

(16) An airplane pilot wishes to fly due north. A wind of 30 km/hr is blowing due west. (a) If the speed of the plane in still air is 150 km/hr, in what direction should the pilot head? (b) What will be his speed as seen by an observer on the ground?

(17) A vector quantity is one that has _____ .

(18) A state of equilibrium exists when _____ .

(19) The following information is known about vectors **A** and **B**. A_x is two units long directed leftward and A_y is four units long directed upward. B_x is six units long directed to the right and B_y is one unit long directed downward. Find the components, magnitude and direction of vectors **C** and **D** if **C** = **A** + **B** and **D** = **A** - **B**.

(20) Express the magnitude of vector **A** in terms of its components A_x and A_y.

(21) A force of 3 N and a force of 8 N act concurrently to yield

a resultant of 11 N. What is the angle between the forces?

(22) What is the effect on the magnitude of the resultant of two forces if the angle between the forces is decreased from 120° to 30°?

(23) The various force vectors acting on an object in equilibrium, when arranged head-to-tail, form _____ .

(24) A vector is 5 units long, directed 30° north of west. Find a set of three components of the given vector. (Hint: Express the given vector and its three components in terms of x and y components.)

(25) What is the effect on the magnitude of the x component of a vector if the vector is rotated from the 30° north of east direction to the 60° north of east direction?

(26) Is it easier or more difficult to push a lawnmower with the handle inclined 30° from the horizontal than at a 60° angle of inclination? Explain your answer.

(27) The resultant of all the forces acting on a body and their equilibrant is _____ .

(28) A 100 N weight is prevented from sliding down a frictionless incline that is oriented 40° from the horizontal. Does the task become easier or more difficult if the angle of inclination is increased to 70°? Explain your answer.

5

CHAPTER FIVE

LAWS OF MOTION

5.1 VOCABULARY

WORD	SYMBOL	UNIT
Force	F	Newton, N
Mass	m	Kilogram, kg
Inertia	m	same as mass
Action, reaction	F	same as force
Weight	W	same as force

DEFINITIONS

LAW OF INERTIA - An object in motion will forever maintain its speed and direction so long as no unbalanced (net) force acts on it.

LAW OF FORCES - The magnitude of the net force acting on an object is equal to the product of the object's mass and its acceleration. $F = ma$. The law is also applicable in vector form, as such: $\mathbf{F} = m\mathbf{a}$.

LAW OF ACTION-REACTION - Every *action* (force) comes

with an equal and opposite *reaction* (also force). Forces come in pairs with the members of each such pair being equal in magnitude and opposite in direction.

FORCE - Loosely defined as a *push* or *pull*. More accurately defined as an influence exerted on an object in an attempt to change the motion status of the object.

MASS - The amount of material (matter, substance, stuff) an object contains.

INERTIA - Tendency of objects to maintain the status quo insofar as motion is concerned. Or, the resistance of an object to change in its motion status. The more mass an object has, the more difficult it is (and the more force is necessary) to change its motion status. Thus, the more mass an object has, the greater its inertia.

NEWTON - Magnitude of force that accelerates a one kilogram object at the rate of one meter per second squared. This is the unit of force in the MKS system.

KILOGRAM - Fundamental unit of mass in the MKS system.

WEIGHT - The force of gravity acting on an object. On earth, at sea level, the formula $W = mg$ is applicable, where g is 9.8 m/s^2.

5.2 THINGS TO KNOW

(1) Force and motion are not synonomous and are not to be confused with each other. An object can be moving, yet no force acts on it. A force may act on an object, yet the object does not move. An object may move one way at the same time that a force (even a net force) acts on it in the opposite direction. For example, an object thrown vertically upward continues to rise for some time

Laws Of Motion 117

while gravity acts to pull it down (the object rises until its speed decelerates to zero). The only relationship between force and motion is that provided by the formula $F = ma$.

(2) A net (resultant) force produces change in motion. Net force is associated with acceleration (or deceleration) in the direction in which the net force acts. In using the formula $F = ma$ (without the vector symbol in the form of a bold **F** and **a**) remember that F represents the magnitude of the *resultant of all the forces* acting on an object (added as vectors) and a represents the magnitude of the acceleration vector. To help remember this it might be a good idea to always write $F_n = ma$ thereby emphasizing the fact that it is *F-net* that is equal to the product *ma*.

(3) A force, defined loosely as a push or pull, can be exerted on an object without making physical contact with the object.

(4) As a book, for example, rests on a table it is influenced by the earth and by the table. The earth's gravitational pull attempts to bring the object closer to the earth's center and the table acts to prevent that from happening. Both, the attraction to the earth and the opposition to it, are exertions to influence the motion status of the book and, as such, constitute forces. The earth's exertion to bring the book closer to itself constitutes an influence (force) that is directed downward; the table's exertion to prevent the book from moving downward constitutes an influence (force) on the book that is directed upward. The only way to counter a downward pull is to push upward. The book remains at rest because the two opposing exertions are equal in magnitude, leading to a state of equilibrium (net force of zero). This delicately balanced arrangement does not occur by accident. It is due to the fact that the upward force exerted by the table is a *response force* - it is only as strong as it need be to

maintain the shape of the table by preventing the book from moving through it. If the book is too heavy or the table too weak, the table may very well collapse under the strain and the force of gravity wins the tug-of-war. The book then takes off and accelerates downward, toward the earth.

(5) *Weight* and *mass* are not the same. They have different meanings, are measured in different units, their values are not the same for any particular object and different instruments must be used to determine their magnitudes. Mass is the amount of material an object contains. It is a scalar quantity, is measured with a balance scale and its magnitude is expressed in kilograms. Weight, on the other hand, is the name of a particular force - the force exerted by gravity. It is a vector quantity (as is any force), is measured with a spring scale and its magnitude is expressed in Newtons. An object's weight is slightly different at the north pole than at the equator (where it is farther from the earth's center) and much different on the moon or on Jupiter than on earth. Yet the object's mass is the same in all these places. An object in deep space may have no weight at all, yet its mass is exactly the same as on earth. Mass and weight just cannot be interchanged!

The mass and weight of an object on earth, at sea level, are related by the formula $W = mg$, where g is 9.8 m/s^2.

(6) The vector characteristics of the formula $\mathbf{F} = m\mathbf{a}$ can easily be expressed in terms of components, as such:

$$\mathbf{F} = m\,\mathbf{a}$$
$$(\mathbf{F_x}, \mathbf{F_y}) = m\,(\mathbf{a_x}, \mathbf{a_y})$$
$$(\mathbf{F_x}, \mathbf{F_y}) = (m\,\mathbf{a_x}, m\,\mathbf{a_y})$$
$$\mathbf{F_x} = m\,\mathbf{a_x}$$
$$\mathbf{F_y} = m\,\mathbf{a_y}$$

Laws Of Motion 119

where the product of a vector and a scalar, such as m and $\mathbf{a}$, is another vector in the same direction whose magnitude is m times the magnitude of the original vector.

The components of vector $\mathbf{a}$, in turn, are defined by the vector equation:

$$\mathbf{a} = \Delta \mathbf{V}/\Delta t$$
$$(\mathbf{a}_x, \mathbf{a}_y) = (\Delta \mathbf{V}_x/\Delta t, \ \Delta \mathbf{V}_y/\Delta t)$$
$$\mathbf{a}_x = \Delta \mathbf{V}_x/\Delta t \quad \text{and} \quad \mathbf{a}_y = \Delta \mathbf{V}_y/\Delta t$$

Dividing by a scalar, such as Δt, is the same as multiplying by the inverse of the scalar.

The bottom line is: the x-component of $\mathbf{F}$ (the net force) is equal to the mass times the x-component of $\mathbf{a}$, and the y-component of $\mathbf{F}$ is equal to the mass times the y-component of $\mathbf{a}$. And the magnitude of $\mathbf{F}$-net (the resultant of all the forces acting on an object) is equal to the mass times the magnitude of $\mathbf{a}$.

(7) Forces come in pairs. Single forces do not occur. The members of each such pair are equal in magnitude and opposite in direction, but don't act on the same object. More specifically, if object A exerts a force on object B, object B returns the favor and exerts an equally strong but oppositely directed force on object A. These forces act at the same time, but since they do not act on the same object they are not concurrent and cannot be added together for any purpose. A corollary of this principle is that an object cannot exert a force on itself.

(8) In tackling complicated problems with many masses and forces it is imperative that each force be associated with the mass or masses upon which it acts. The formula $\mathbf{F} = m\mathbf{a}$ relates the resultant force to the acceleration it produces in the object *upon which the resultant force acts*.

5.3 SOLVING THE PROBLEMS

(1) A rocket has a mass of 10,000 kg. How strong must the upwardly directed force of propulsion be in order that the rocket accelerate upward at the rate of 4 m/s^2 ?

SKETCH AND IDENTIFY

The force of propulsion is not the only force acting on the rocket. The ever present pull of gravity continues to play its role. Thus, the net force $\mathbf{F_n}$ on the rocket is the vector sum of the upward acting force of propulsion, $\mathbf{P}$, and the downward acting force of gravity, $\mathbf{W}$. The magnitude of the resultant of these forces, F_n (no vector symbol), is the difference between the magnitude of $\mathbf{P}$ and the magnitude of $\mathbf{W}$. This is illustrated in figure 5-1 via the *head to tail* method of vector addition. We know that $\mathbf{P}$ must be longer than $\mathbf{W}$ since the rocket accelerates upward and the acceleration vector is always oriented in the same direction as the net force vector.

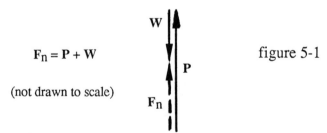

$\mathbf{F_n} = \mathbf{P} + \mathbf{W}$

(not drawn to scale)

figure 5-1

LINK UNKNOWN TO GIVEN

$\qquad\qquad F_n = m\,a \qquad$ (magnitude only, no vector symbol)

$\qquad P - W = m\,a$

$\qquad P - mg = m\,a \qquad$ (since $W = mg$)

$P - (10,000)(9.8) = (10,000)(4)$

$\qquad\qquad P = 40,000 + 98,000 = 138,000$ N

Laws Of Motion

(2) A 10 kg shell is fired from the barrel of a horizontal artillery gun. The shell emerges from the 2 meter long barrel with a speed of 500 m/s. How strong was the average force exerted on the shell inside the barrel?

SKETCH AND IDENTIFY

The shell's trip through the barrel begins at one end with zero velocity and ends at the other end with a velocity of 500 m/s. The average acceleration inside the barrel can be determined by applying the motion formulas of chapter two. The average force is related to the average acceleration by the formula $F = ma$.

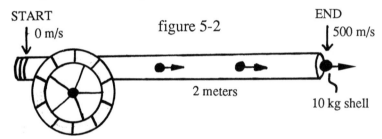

figure 5-2

LINK UNKNOWN TO GIVEN

$v_f^2 - v_i^2 = 2ad$ (trip across barrel)

$500^2 - 0^2 = 2a(2)$

$250000 = 4a$

$a = 62{,}500 \text{ m/s}^2$ (average acceleration)

Only one force is known to exist in this case, the force exerted by the exploding gunpowder on the shell. Thus,

$F = ma$

$F = (10)(62500)$

$F = 625{,}000 \text{ N}$ (average force)

(3) A 5,000 kg car is moving at the rate of 80 km/hr when the

driver suddenly becomes aware of a pedestrian crossing the street. The pedestrian is 20 meters from the car at the instant the driver applies the brakes. A 200 kg passenger is seat-belted into the front seat, near the driver. (a) How strong must the braking force be in order that the car is stopped before colliding with the pedestrian? (b) How much tension must the seat belts be able to exert in order that they succeed in keeping the passenger tied to his seat ? Assume no friction between the passenger and his seat.

SKETCH AND IDENTIFY

The passenger is, of course, moving together with the car at 80 km/hr when the brakes are applied. But the brakes act on the car, not on objects inside the car. If the passenger were not belted to the seat, he would continue moving at 80 km/hr during the time that the car is decelerated by the brakes. The passenger would then be traveling faster than the car and head straight into the windshield. (Friction between the passenger and his seat might act to slow him down somewhat, but at this speed would not suffice to prevent him from sliding off the seat. In any event, we are told to assume that no friction exists.)

This is where seat belts come in. They act to bind the passenger to the car. As the brakes are applied and the passenger begins to move forward relative to the car, his body acts to pull the belt toward the front of the car. The belt, in turn, fights back because it resists being stretched. It exerts a force on the passenger directed opposite to the motion of the car, in an attempt to prevent the passenger from moving forward relative to the car. In other words, the seat belts act to keep the passenger in his seat by pulling on him in the backward direction. For the passenger to remain in his seat he

Laws Of Motion

must at all times be moving at the same speed as the car. This then becomes the objective of the tension force exerted by the seat belts on the passenger - to keep him moving together with the car (rather than take off at the rate of 80 km/hr). In other words, the seat belts act to decelerate the passenger at the same rate that the brakes act to decelerate the car. The passenger must be slowed to a stop in the same time and distance as the car takes to be brought to a stop. This is illustrated in figure 5-3 where the single arrows represent motion and the double arrows represent force.

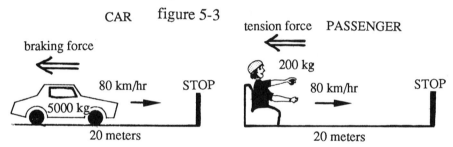

figure 5-3

LINK UNKNOWN TO GIVEN

To find the deceleration rate for both, the car and the passenger, we proceed as follows:

$v_f^2 - v_i^2 = 2ad$ (from application of brakes to stopping)

$0^2 - 22.2^2 = 2a(20)$ (converting 80 km/hr to 22.2 m/s)

$a = -12.32 \text{ m/s}^2$

The braking force on the car can be determined as follows:

$F_n = m\,a$ (magnitude only)

Since the braking force F_b is the only force acting on the car, it is the net force. Thus:

$F_b = (5000)(12.32) = 61{,}600 \text{ N}$

The tension force F_T on the passenger can analogously be determined, as follows:

$$F_n = m\,a$$
$$F_T = m\,a$$
$$F_T = (200)(12.32) = 2464 \text{ N}$$

(4) (a) What is the acceleration rate of the system illustrated below (figure 5-4)? Assume no friction between the 10 kg mass and the table or between the string and the wheel, and ignore the mass of the string and wheel. (b) How strong is the tension in the string?

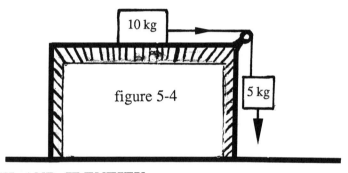

figure 5-4

SKETCH AND IDENTIFY

The 5 kg mass hanging over the edge of the table is subjected to two forces: its own weight W_1 acting downward and the force of tension T_1 exerted on it by the string in the upward direction. As gravity acts to pull the 5 kg mass downward, the string is pulled downward too. But the 10 kg mass at the other end resists being disturbed from its state of rest - due to its own inertia. To avoid being stretched the string acts to overcome this resistance by exerting a force on the 10 kg object to get it moving. But every action (force) is associated with an equal and opposite reaction. The 10 kg mass responds by pulling on the string in the opposite direction - to the left. The net result is that the two masses act to pull the string in opposite directions. The string resists being stretched and fights back by pulling both objects inward - toward its center. This means that the string acts to pull the 5 kg mass upward and the 10 kg mass to the right (figure 5-5).

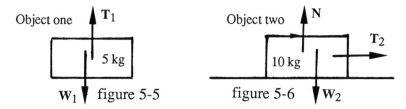

The 10 kg mass is subjected to three forces: its own weight W_2 acting vertically downward, a normal force N exerted by the table in the upward direction, and the tension T_2 exerted by the string to the right (figure 5-6). The oppositely directed normal and weight forces acting on the 10 kg mass must be equal in magnitude since the 10 kg object is not accelerated upward or downward, but to the right. The resultant force on the 10 kg mass can therefore be due only to T_2. The oppositely directed tension and weight forces acting on the 5 kg mass, on the other hand, are not equal in magnitude since the 5 kg mass is accelerating downward, in the direction of its weight. The net force on the 5 kg object is $W_1 + T_1$. Since these forces are directed opposite to each other, the magnitude of their resultant is $(W_1 - T_1)$.

Now, the magnitude of the tension forces acting, respectively, on the 5 and 10 kg objects must be equal to each other. The upward force exerted by the string on the 5 kg object, T_1, and the rightward force exerted by the string on the 10 kg object, T_2, are both consequences of the same act - the string's resistance to being stretched. And since the two masses can only move in tandem (the length of the string is kept fixed) their respective speeds must also, at any point in time, be equal to each other. Therefore, any acceleration rate experienced by one must be experienced by the other. The acceleration rates of the two masses must therefore also be equal to each other. The bottom line is that as far as magnitudes are concerned there is only one tension value T and one acceleration rate a in this problem.

LINK UNKNOWN TO GIVEN

We apply F = ma separately to each object, named one and two, using the appropriate force and mass in each case.

For the 5 kg mass (object *one*):

$F_n = m\,a$ (magnitude only, no vector symbol)

$W_1 - T = m_1\,a$

$m_1\,g - T = m_1\,a$ (W = mg for any object at sea level)

$(5)(9.8) - T = 5\,a$

For the 10 kg mass (object *two*):

$F_n = m\,a$ (magnitude only)

$T = m_2\,a$

$T = 10\,a$

Now we replace T in the equation above with *10a* and obtain:

$49 - 10\,a = 5\,a$

$49 = 15\,a$

$a = 3.27 \text{ m/s}^2$ (acceleration rate for both objects)

$T = (10)(3.27) = 32.7 \text{ N}$ (both objects)

(5) A 100 kg mountain climber is to be lowered down the side of a cliff with a rope. The maximum tension the rope can withstand is 600 N. (a) Can the mountain climber safely be lowered at a slow

Laws Of Motion

and steady rate? (b) If not, what ought to be done to make sure the rope does not snap?

SKETCH AND IDENTIFY

As the mountain climber is lowered she is subjected to two forces: her weight **W** acting downward and the force of tension **T** acting upward. The magnitude of the resultant of these forces is the difference between their magnitudes (figure 5-7). In other words, $F_n = W - T$. By plugging this net force into the formula $F = ma$ we can determine the magnitude of the tension force, T, for various acceleration rates a. A procedure for lowering the mountain climber can be deemed safe only if the tension is less than the maximum the rope can withstand.

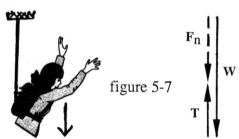

figure 5-7

LINK UNKNOWN TO GIVEN

To determine the magnitude of **T** as the mountain climber is lowered at a slow and steady pace, we proceed as follows:

$F_n = m\,a$ (magnitude only, no vector symbol)

$W - T = (100)(0)$ (constant velocity, $a = 0$)

$T = W = mg$

$T = (100)(9.8) = 980 \text{ N}$

This exceeds the maximum tension the rope can tolerate. Lowering the mountain climber in a slow and steady manner is therefore not safe - the rope may snap.

To decrease the tension in the rope the mountain climber should be lowered in a manner that allows her to accelerate downward. The net force then is directed downward, W is greater than T due to a reduction in the tension, and the physics appears as follows:

W - T = m a   (magnitude only)

T = W - ma

To make sure the rope does not snap the force of tension should be equal to or less than 600 N (the maximum allowed).

$$T \leq 600$$
$$W - ma \leq 600$$
$$mg - ma \leq 600 \qquad \text{(since } W = mg\text{)}$$
$$mg - 600 \leq ma$$
$$(100)(9.8) - 600 \leq 100\,a$$
$$380 \leq 100\,a$$
$$a \geq 3.8 \text{ m/s}^2$$

The mountain climber must be lowered with a downward acceleration rate of *at least* 3.8 m/s². The greater the downward acceleration rate, the smaller the tension in the rope and the safer the procedure. At a downward acceleration rate of 4 m/s², for example, the equation *T = W - ma* becomes *T* = 980 - 400, or 580 N, which is perfectly safe since it is below the maximum of 600 N.

(6) A 200 N man stands on a scale in an elevator. What is the scale reading when the elevator is (a) at rest, (b) moving between floors at the constant speed of 4 m/s, (c) accelerating upward at the rate of 2 m/s², (d) accelerating downward at the same rate, (e) falling freely after the cable snaps? Assume no friction between the elevator and the walls of the shaft.

Laws Of Motion

SKETCH AND IDENTIFY

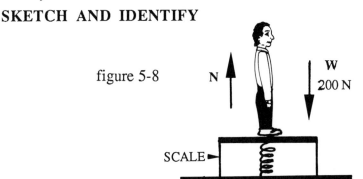

figure 5-8

A man in an elevator is acted upon by two forces: his own weight **W** directed downward and the normal force **N** exerted on him by the floor of the elevator upward. The normal force stems from the floor's rigidity which resists compression and deformation. It acts to oppose gravity's attempt to draw the man closer to earth.

The scale can be viewed as a type of floor (the spring resists compression just as a solid floor does) with an additional, advantageous feature - it provides a reading of the magnitude of the force exerted downward on it. Since every action is associated with an equal and opposite reaction, the downward force exerted by the man on the scale is equal in magnitude to that exerted by the scale on the man upward. The net result is that the scale reading is equal to the normal force.

We are accustomed to think of a scale reading as an indication of a person's weight. This is true under normal circumstances, with stationary people on stationary floors in stationary rooms. Then the person standing on the scale is in a state of equilibrium and the normal force (the scale reading) is equal and opposite to the weight (as is the situation in part *a* of the problem). It is not true as a rule, however. The weight of a person is the force exerted by the earth on him, not the force he exerts on the scale. The scale, on the other hand, indicates the force exerted on it. If for some reason these two quantities are not equal (as is the case in parts *c, d* and *e*) the scale reading represents the force exerted on it, not the person's weight. And the force exerted on the scale by the person standing on it is equal in magnitude to the normal force exerted by the scale on the person, since these two forces constitute an action-reaction pair. (The reaction to the earth's pull on the person - his weight - is the force exerted by the person on the whole earth upward. This force does not enter this problem, however, since we are interested in the forces acting on the person.)

Why, you may wonder, should the force exerted downward by the person on the scale ever be different from the person's weight? For many possible reasons. For example, when the elevator accelerates upward the scale must be pushing the man upward with more force than the person's weight in order to overcome gravity and achieve a net force in the upward direction. The man, in turn, *must* respond to this by pushing down harder on the scale due to the principle of action-reaction.

The weight of the man in our problem does not depend on what he or the elevator is doing. It is fixed at 200 N. But the normal force does depend on what the elevator is doing, as does the scale reading.

We apply the formula F = *m*a to the man in the elevator, making sure that the correct *F*- net, *m* and *a* are employed in each case.

LINK UNKNOWN TO GIVEN

(a) Elevator at rest:

$F_n = m\,a$ (magnitude only, no vector symbol)

$W - N = m\,a$

Laws Of Motion

$200 - N = (200/9.8)(0)$ (Since $W = mg$, $m = W/g$)
$N = 200$

The scale reading equals the weight of 200 N.

(b) Constant speed, 4 m/s:

Same as part *a*, since the acceleration rate is still zero. The scale reading equals the weight of 200 N.

(c) Accelerating upward, 2 m/s^2:

The magnitude of **N** must be greater than **W** in this case. Let's keep F-net positive by subtracting the magnitude of the smaller force, W, from the magnitude of the larger force, N.

$N - W = m\,a$
$N - 200 = (200/9.8)(2)$ $(m = W/g)$
$N = 240.8$ Newtons

The scale reads 240.8 N, a number that is larger than the weight.

(d) Accelerating downward, 2 m/s^2:

$W - N = m\,a$ (W larger than N)
$200 - N = (20.4)(2)$
$N = 159.2$ Newtons

The scale reads 159.2 N, a number that is smaller than the weight. The elevator floor now only partially opposes gravity, allowing the person to fall with some acceleration downward (but less than the acceleration rate of free fall - 9.8 m/s^2).

(e) Free fall:

$W - N = m\,a$ (W larger than N)
$200 - N = (20.4)(9.8)$
$N = 0$

The scale reads zero. The elevator floor is now not pushing upward on the person at all, allowing gravity to accelerate him freely toward the earth.

(7) A block slides down a smooth, frictionless incline. The angle of inclination is 30°. What is the acceleration rate of the block?

SKETCH AND IDENTIFY

Two forces act on the block in this problem: the pull of gravity, **W**, directed vertically downward and the normal force, **N**, exerted by the incline perpendicularly to the surface of the incline (figure 5-9). Let us think our way through these forces to their resultant. Vector **W** can be replaced by its two perpendicular components, one

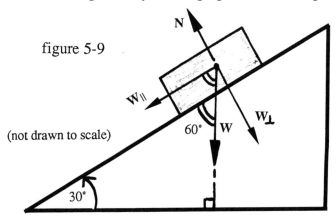

figure 5-9

(not drawn to scale)

parallel to the incline, $W_{\parallel}$, and one perpendicular to the incline, $W_{\perp}$. The normal force, **N**, must be equal and opposite to $W_{\perp}$ since the resultant force can have no component in the direction perpendicular to the incline - it must be entirely directed parallel to the incline. So when we add our three vectors, $W_{\parallel}$, $W_{\perp}$ and **N**, to find their resultant, we are left only with $W_{\parallel}$. The magnitude of the resultant is therefore equal to the magnitude of $W_{\parallel}$.

Since we are not provided with the mass or weight of the block, it can only be assumed that these factors do not matter. In other words, the acceleration rate down a smooth, frictionless incline is

Laws Of Motion

independent of the mass or weight of the block. This implies that any *m* or *W* in the equations we develop will drop out by the time we solve for *a*.

LINK UNKNOWN TO GIVEN

The magnitude of $W_{\|}$ is determined by vector resolution, as follows (figure 5-10):

$\cos 60° = W_{\|}/W = .5$ (magnitude only)

$W_{\|} = .5\, W$

$F_n = m\, a$ (magnitude only)

$W_{\|} = m\, a$

$.5W = m\, a$

$.5mg = m\, a$ $\quad (W = mg)$

$a = .5g = (.5)(9.8) = 4.9 \text{ m/s}^2$

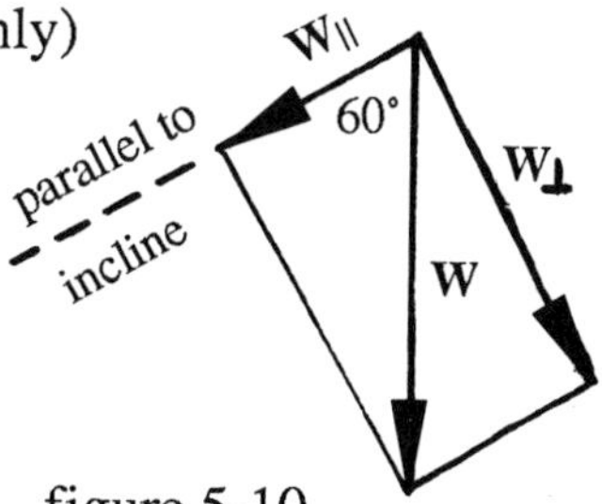

figure 5-10

(8) A 4 kg object is moving due east at the rate of 20 m/s. A force of 8 N directed 60° north of west then acts on the object for 5 seconds. What is the magnitude and direction of the object's velocity after the five seconds are over?

SKETCH AND IDENTIFY

The force will accelerate the object in the direction in which it acts. This acceleration imparts a new velocity vector, ΔV, to the object which is to be added to the original velocity vector, V_i, to yield the resultant velocity vector, V_f, after the five second interval is over.

We can proceed either by components or via the other rules of vector addition. Let's solve the problem both ways.

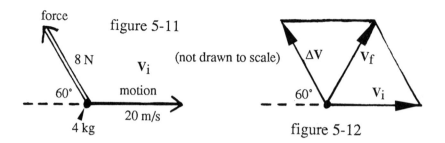

figure 5-11 (not drawn to scale)

figure 5-12

LINK UNKNOWN TO GIVEN

BY COMPONENTS

$\mathbf{F} = (\mathbf{F_x}, \mathbf{F_y})$

The magnitudes of the components, F_x and F_y, are determined by vector resolution as follows (figure 5-13).

$\cos 60° = F_x/F = .5 \quad \sin 60° = F_y/F = .866$

$F_x = (8)(.5) \qquad\qquad F_y = (8)(.866)$

$F_x = 4 \text{ N} \qquad\qquad F_y = 6.93 \text{ N}$

Now, we know that $\mathbf{F_x} = m\mathbf{a_x}$ and $\mathbf{F_y} = m\mathbf{a_y}$ (Things to Know, item six). This implies that the magnitude of $\mathbf{F_x}$ is equal to m times the magnitude of $\mathbf{a_x}$, and the magnitude of $\mathbf{F_y}$ is equal to m times the magnitude of $\mathbf{a_y}$.

(not drawn to scale)

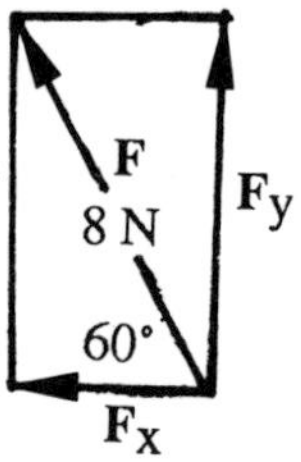

$F_x = m\, a_x \qquad\qquad F_y = m\, a_y$

$4 = 4\, a_x \qquad\qquad 6.93 = 4\, a_y$

$a_x = 1 \text{ m/s}^2 \qquad a_y = 1.73 \text{ m/s}^2$

$a_x = \Delta V_x / \Delta t \qquad a_y = \Delta V_y / \Delta t$

$1 = \Delta V_x / 5 \qquad\quad 1.73 = \Delta V_y / 5$

$\Delta V_x = 5 \text{ m/s} \qquad\quad \Delta V_y = 8.66 \text{ m/s}$

figure 5-13

Laws Of Motion 135

These are the components of the ΔV vector - the newly imparted velocity vector which must be added vectorially to the original velocity vector (before the force appeared on the scene) to arrive at the final (resultant) velocity vector of the object (after the force is removed).

$$V_f = V_i + \Delta V$$
$$(V_{fx}, V_{fy}) = (V_{ix}, V_{iy}) + (\Delta V_x, \Delta V_y)$$
$$V_{fx} = V_{ix} + \Delta V_x \quad \text{and} \quad V_{fy} = V_{iy} + \Delta V_y$$

Since the vectors to be added on the right side of each of these equations are in the same or in opposite directions (they are all either *x* or *y* component vectors) we can find the magnitude of their resultant by adding those directed one way (rightward or upward), then subtracting those directed the opposite way (leftward or downward). That magnitude is, in turn, equal to the magnitude of the vector that appears on the left side of each of these equations. Thus:

$$V_{fx} = 20 - 5 = 15 \text{ m/s} \quad \text{and} \quad V_{fy} = 0 + 8.66 = 8.66 \text{ m/s}$$

The object's final velocity vector has components 15 m/s and 8.66 m/s in the *x* and *y* directions, respectively (figure 5-14). The magnitude of this velocity vector is determined as follows:

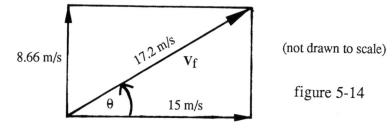

(not drawn to scale)

figure 5-14

$$V^2 = 15^2 + 8.66^2$$
$$V^2 = 300$$
$$V = 17.2 \text{ m/s}$$

The direction of the final velocity is determined as such:

Tan θ = 8.66/15 = .577

θ = 30°

The final velocity vector is directed 30° north of east.

ALTERNATE METHOD

$F_n = ma$ (magnitude only, applied in direction of force)

8 = 4 a

a = 2 m/s²

a = $\Delta V/\Delta t$

2 = $\Delta V/5$

ΔV = 10 m/s

Since we know that ΔV is directed 60° north of west (same direction as the force) and the final velocity vector is the resultant of the original velocity vector and the new ΔV vector, we proceed as in figure 5-15 where the problem is solved by construction.

SCALE: 1 cm = 4 m/s

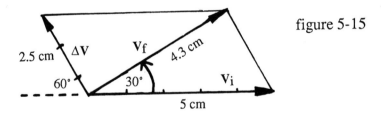

figure 5-15

The magnitude of the resultant velocity vector turns out (by measurement) to be 4.3 cm long, representing a speed of 17.2 m/s (based on the chosen scale) and the direction of the resultant turns out (also by measurement) to be 30° N of E.

5.4 ON YOUR OWN PROBLEMS

(Asterisk indicates solution appears in appendices.)

*(1) A 5000 kg automobile is coasting at the rate of 50 km/hr. The driver then presses down on the gas pedal for 20 seconds, after which the speedometer indicates that the car is moving at the rate of 90 km/hr. How strong was the force applied by the engine to the car? Assume no friction.

*(2) A 10 gram bullet moving at the rate of 100 m/s is aimed at a tree. The bullet penetrates 12 cm into the tree and becomes embedded there. How strong a force did the bullet exert on the tree?

*(3) Determine the acceleration of the system illustrated in figure 5-16. Assume no friction and disregard the mass of the string and wheel.

Figure 5-16

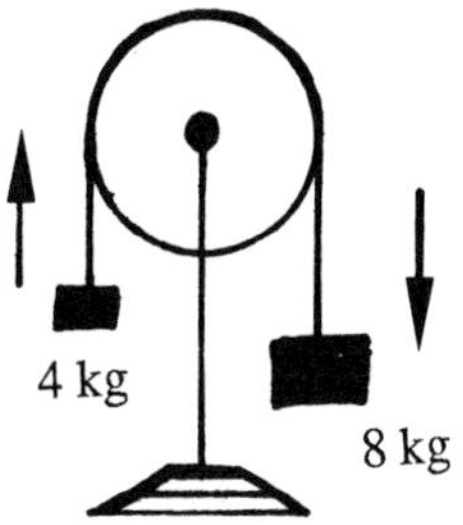

4 kg

8 kg

*(4) A 20 kg mass is pulled and accelerated upward by a rope. The maximum tension the rope can withstand is 500 N. What is the maximum acceleration that can be given to the mass without causing the rope to snap?

*(5) An upward acting force of 200 N is applied to the system illustrated in figure 5-17. Assume no friction and ignore the mass of the string. (a) How strong is the tension in the string? (b) What is the acceleration of the system?

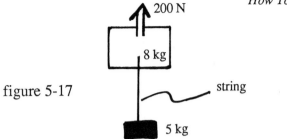

figure 5-17

*(6) A crate sits on a spring scale that is suspended from the ceiling of an elevator. (a) If the scale reads 60 N when the elevator is accelerating upward at the rate of 4 m/s², what is the mass of the crate? (b) How is the elevator moving when the scale reads 8 N?

*(7) A block is launched up an incline with a velocity of 40 m/s. The angle of inclination is 60°. How far up the incline will the block travel before turning back? Assume the incline to be smooth and frictionless.

*(8) A 100,000 kg rocket is lifted vertically upward off its launching pad with an acceleration rate of 5 m/s². How strong is the force of propulsion acting on the rocket?

(9) A 2000 kg motorcycle accelerating uniformly from rest travels a distance of 300 meters in 10 seconds. How strong is the force exerted on the motorcycle by its engine? Assume no friction.

(10) A force of 50 N is applied to a mass of 200 kg. If the mass starts from rest, how far will it travel in 30 seconds?

(11) A baseball moving at the rate of 30 m/s pushes the catcher's hand backward a distance of 20 cm before coming to a stop. How strong a force did the ball apply to the hand?

(12) If the mass of the object on the right side of the wheel in problem three is doubled, what will be the acceleration rate of the system?

(13) A 100 kg crate is lowered from the roof of a building by a

Laws Of Motion

rope. The maximum tension the rope can withstand is 300 N. What minimum acceleration rate must the crate be lowered with in order that the rope does not snap?

(14) When is the tension in an elevator cable the greatest - when the elevator is going up at maximum speed, going down at maximum speed, accelerating upward, accelerating downward, or at rest? Explain your answer.

(15) A force is applied to the system in figure 5-18 causing it to accelerate horizontally at the rate of 20 m/s². (a) What is the tension in the string? (b) How strong is the force acting on the system? Assume no friction between the two objects and the surface over which they are situated.

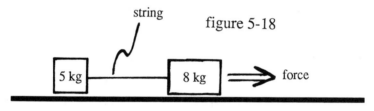

figure 5-18

(16) A block slides down a smooth, frictionless incline. The angle of inclination is 30° and the block starts from rest. What is the speed of the block after it slides thru a distance of 4 meters?

(17) A 40,000 kg spaceship far from earth is decelerated by the firing of retro-rockets at the rate of 3 m/s². How strong is the force applied to the spaceship?

(18) A truck weighing 5000 N is moving at the rate of 30 m/s. The brakes are applied and the truck comes to a stop in 200 meters. Assuming the engine was inactive (foot off the gas pedal) while the brakes were applied, and ignoring friction, how strong a force did the brakes apply to the truck?

(19) How many times as strong would the braking force in

problem eighteen need to be in order to stop the truck in the same distance if it is moving twice as fast?

(20) A 6 kg object is moving due west at the rate of 20 m/s. A force of 3 N directed due south is then applied for 30 seconds. What is the magnitude and direction of the object's velocity when the force ceases to act?

(21) An astronaut in space suddenly notices a 50 kg asteroid moving directly toward him at the rate of 20 m/s. The astronaut takes out his laser gun and aims it at the asteroid when the asteroid is 200 meters from him. How strong must be the force applied by the laser beam to the asteroid in order that the asteroid is stopped before it collides with the astronaut?

(22) If avery action is associated with an equal and opposite reaction it follows that the pull of a horse on a wagon leads to an equal and opposite pull. Does this not imply that the net force is zero? If the net force is zero the acceleration rate should be zero, since $F_n = ma$. How then does any horse make a wagon move?

(23) A 5 kg object is taken to Mars and dropped from a height of 20 meters on the martian surface. The object is then observed to fall freely, taking 6 seconds to arrive on the ground. Assuming it accelerated uniformly as it fell, what is the object's weight on Mars?

6

CHAPTER SIX

FRICTION

6.1 VOCABULARY

WORD	SYMBOL	UNIT
Friction	$\mathbf{F}_f$	Newton, N
Normal	**N**	Newton, N
Coefficient of friction	μ_k, μ_s	none

DEFINITIONS

FRICTION (kinetic) - Force opposed to motion that is associated with the rubbing of one surface against another.

FRICTION (static) - Force exerted on an object at rest to prevent motion that would necessitate the rubbing of one surface against another.

NORMAL - Force exerted by one object on another directed perpendicularly to the surface of contact between the objects.

COEFFICIENT OF KINETIC FRICTION (μ_k) - Number assigned to a pair of surfaces equal to the ratio of the magnitude of the force of friction to the magnitude of the normal force in the case where relative motion between the surfaces (in contact with each other) exists. Different surfaces generally have different coefficients of kinetic friction. $F_f = \mu_k N$.

COEFFICIENT OF STATIC FRICTION (μ_s) - Number assigned to a pair of surfaces equal to the ratio of the maximum possible force of friction to the magnitude of the normal force in the case where the surfaces in contact are at rest relative to each other. Different surfaces generally have different coefficients of static friction. $F_f \leq \mu_s N$.

6.2 THINGS TO KNOW

(1) The normal force, **N**, is the force exerted by either object on the other perpendicularly to the suface of contact between the objects. Frequently it is equal to the weight of one of the objects, but many times it is not. As a rule, the normal force, **N**, is *not* synonomous with weight, **W**, and the two quantities ought not be confused with each other.

(2) The magnitude of the force of friction, F_f, present under static conditions is *not* necessarily equal to $\mu_s N$. The formula $F_f \leq \mu_s N$ provides the maximum force of friction that can be exerted to prevent relative motion between two surfaces in contact. If friction can accomplish this goal (of preventing relative motion) with less force than $\mu_s N$, it exerts the lesser force. In other words, the force of friction under static conditions is, like the forces of tension and compression, a *response* force - it is only as strong as it need be to

Friction   143

achieve its goal (see section 5.2).

As an example, consider the case of a book at rest on a table. If no effort is made to get the book moving, the force of static friction present is zero - although the quantity $\mu_s N$ is equal to, say, 20 N. If a 10 N force is exerted by someone in an attempt to get the book moving, the force of static friction present is only 10 N strong (not 20 N) since that is all the force needed to keep the book at rest. The force of friction is always directed opposite to motion or attempted motion, and a 10 N force of friction suffices in this case to produce equilibrium. The object begins to move only when the applied force, F_A, exceeds the quantity $\mu_s N$. The magnitude of the net force on the object is then equal to $F_A - \mu_s N$, the applied force minus the maximum possible opposing force of static friction.

(3) The above (item two) is not applicable to kinetic friction. The force of friction under kinetic conditions is always *equal to* $\mu_k N$, so long as relative motion between the surfaces in contact continues to exist.

(4) The force of friction in the case of a solid object that slides or attempts to slide over another solid object, is generally independent of speed or the size of the contact area. It depends only on the roughness of the surfaces (represented by the value of μ) and the magnitude of the normal force (N).

6.3 SOLVING THE PROBLEMS

(1) A 200 kg box is dragged along a floor by a rope that makes a 30° angle with the horizontal (figure 6-1). The tension in the rope remains at 500 N as the box continues to move at the constant rate of

5 m/s. What is the coefficient of kinetic friction between the box and the floor?

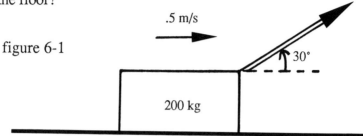

figure 6-1

SKETCH AND IDENTIFY

Since the box moves at constant speed it is neither accelerated nor decelerated. This means that no net force is exerted on it - the box is in a state of equilibrium. The forces exerted in the horizontal dimension add up (as vectors) to zero, and the forces exerted in the vertical dimension add up to zero.

In the horizontal dimension two forces act on the box. There is the x-component of the tension, T_x, acting in the direction of the motion of the box and there is the force of friction, F_f, fighting this motion. Since the force of kinetic friction is always directed opposite to motion, it is also in this case directed opposite to T_x, as illustrated in figure 6-2.

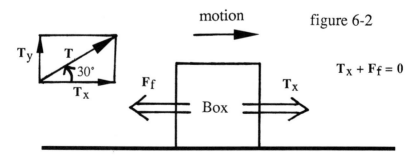

In the vertical dimension there are three forces acting on the box: the y-component of the tension, T_y, directed upward; the force of

Friction

gravity, **W**, directed vertically downward; and the normal force, **N**, exerted by the floor on the box in the upward direction (figure 6-3). The magnitude of the normal force may turn out to be zero - if T_y and **W** are equal in magnitude - but that is not known for a fact at this point.

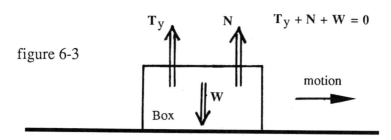

figure 6-3

Since no net force is present in the horizontal dimension it must be true that

$T_x = F_f$ (magnitude only, no vector symbol)

and since no net force is present in the vertical dimension it must be true that

$T_y + N = W$ (magnitude only)

It may appear at first glance that five unknowns are contained in these two equations. But appearances are sometimes deceiving. Since T_x and T_y are components of **T**, their magnitudes can be expressed in terms of the magnitude of **T**. The magnitude of the weight, W, can be determined from the known mass via the formula $W = mg$ and F_f and N are related by the formula $F_f = \mu_k N$. So we really have only two unknowns here. Notice that the magnitude of the normal force in this problem is *not* equal to the weight but to the difference between the weight and the vertical component of the tension ($W - T_y$).

By vector resolution (figure 6-2) we know that:

$$T_x = T \cos 30° \quad \text{and} \quad T_y = T \sin 30°$$

We thus proceed as follows:

$$T_x = F_f \quad \text{(magnitude only)}$$
$$T \cos 30° = \mu_k N$$
$$T_y + N = W$$
$$T \sin 30° + N = mg$$
$$N = mg - T \sin 30°$$
$$T \cos 30° = \mu_k (mg - T \sin 30°)$$
$$(500)(.866) = \mu_k ([200][9.8] - [500][.5])$$
$$433 = \mu_k (1710)$$
$$\mu_k = .25$$

(2) A package falls off a moving truck on a highway. After some detective work it is determined that the package came to a stop 200 meters from the point where it first hit the road. Experiments reveal that the coefficient of kinetic friction between the package and the road is (.6). How fast was the truck moving when the package fell?

SKETCH AND IDENTIFY

An instant before the package fell off the truck its horizontal velocity (parallel to the road) must have been the same as the truck's. This is so since the package was moving with the truck just before it fell. Now, the act of falling (being pulled down vertically by gravity) only effects the vertical position and velocity of a falling object. As the package fell to the ground it must have, therefore, continued to move horizontally in a manner identical to the truck,

Friction

with the same speed, in the same direction. We know of no horizontally directed force acting to effect the horizontal motion of the falling package (before it makes contact with the ground). Finding the horizontal speed with which the package begins its slide on the road is therefore akin to finding the speed of the truck at the instant the package fell.

After the package hits the ground with some horizontal speed the only force to have a discernible effect on it is due to kinetic friction. The weight, **W**, of the package is negated by an equal and opposite normal force **N** exerted upward by the ground, since the package is neither accelerated upward nor downward after it encounters the ground. No acceleration means no net force in the vertical dimension. In the horizontal dimension, however, the package *is* decelerated to a stop, within a distance of 200 meters. If it begins its slide on the road by moving, for example, to the right, the force of kinetic friction acts in the opposite direction, to the left, to oppose this motion (figure 6-4). No force (such as that due to wind action) acts to maintain the motion of the package, as far as we can tell from the problem. So the force of kinetic friction is the net force in this problem.

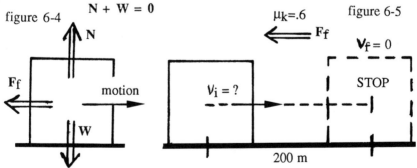

Since the mass of the package is not provided in the problem we can only assume that all *m*'s drop out of the equations at some point.

This happens because the solution to the problem is independent of the mass of the package.

LINK UNKNOWN TO GIVEN

$F_f = \mu_k N$ $N = W$ $W = mg$ (magnitude only)

$F_f = \mu_k mg$

$F_{NET} = ma$ $F_{NET} = F_f$

$F_f = ma$

$\mu_k mg = ma$

$\mu_k g = a$ (mass drops out)

$a = (.6)(9.8) = 5.88$ m/s²   (deceleration rate)

$v_f^2 - v_i^2 = 2ad$ (from hitting the ground to stopping)

$0 - v_i^2 = 2(-5.88)(200)$

$v_i^2 = 2352$

$v_i = 48.5$ m/s

The package landed on the ground with a horizontal velocity of 48.5 m/s. This, therefore, must have been the truck's speed as the package fell - 175 km/hr!

(3) What minimum acceleration must be imparted to the cart in figure 6-6 in order that the block in front of the cart and in contact with it is prevented from sliding down? The coefficient of static friction between the block and the cart is (.4).

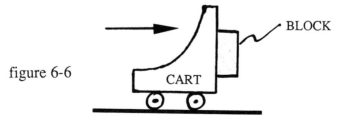

figure 6-6

Friction

SKETCH AND IDENTIFY

As the cart is accelerated, it acts to accelerate the block in front of it. This means that the cart exerts a force on the block directed to the right in figure 6-7. Since this force is exerted by one object on another perpendicularly to the surface of contact between them, it becomes our normal force, **N**. Its magnitude is altogether independent of the block's weight, **W**, which acts vertically downward. Acting to prevent the block from sliding down (under the influence of the pull of gravity) is the force of static friction, $\mathbf{F_f}$, which opposes the downward slide by acting vertically upward (the only way to oppose downward motion is to act upward). The block, whose mass m we do not know, is therefore subjected to three forces, as illustrated in figure 6-7.

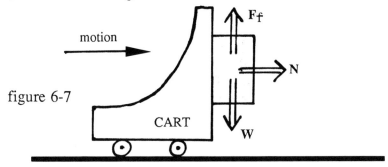

figure 6-7

To be successful in its effort to prevent the block from sliding down, the force of static friction must be equal in magnitude to the weight. Only then will the net force in the vertical dimension be zero. Static friction is, as we know, a *response* force. In this case it responds to the weight by exerting an equally strong but oppositely directed force on the block - assuming this is within its ability to do (the maximum is not exceeded). We insist, therefore, that:

$$F_f = W = mg \qquad \text{(magnitude only)}$$

Now, the force of static friction can, at most, be equal to $\mu_s N$

and N in this case is equal to *ma*, where *a* is the acceleration rate of the block and the cart (they must be the same) to the right.

$$F_f \leq \mu_s N \qquad N = ma \qquad \text{(magnitude only)}$$

LINK UNKNOWN TO GIVEN

$$F_f \leq \mu_s N$$
$$F_f \leq \mu_s ma$$
$$mg \leq \mu_s ma$$
$$g \leq \mu_s a \qquad \text{(mass drops out)}$$
$$a \geq g/\mu_s$$
$$a \geq 9.8/.4$$
$$a \geq 24.5 \text{ m/s}^2$$

(4) Two 10-kg masses hang on either side of an incline, as illustrated in figure 6-8. The coefficient of static friction between the mass that rests on the incline (m_2 in the figure) and the incline is (.4). Neither mass succeeds in pulling itself down and the other up; both masses are at rest. (a) How strong is the tension in the rope? (b) How strong is the force of friction that acts on the mass that rests on the incline?

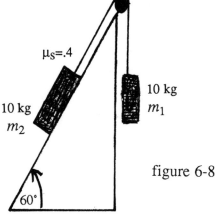

figure 6-8

Friction   151

SKETCH AND IDENTIFY

Four forces act on the mass that rests on the incline (mass-2). The rope exerts a force of tension, **T**$_2$, directed parallel to and toward the top of the incline. The force of gravity, **W**$_2$, acts vertically downward, a component of which acts parallel to and toward the bottom of the incline (**W**$_\parallel$) and a component acts perpendicularly to the incline (**W**$_\perp$). These two components may substitute for **W**$_2$, since their resultant is **W**$_2$. A normal force, **N**, is exerted by the incline on the mass, perpendicularly to the incline and opposite to **W**$_\perp$. And the force of static friction, **F**$_f$, acts to prevent motion over the incline. We cannot, at this point, determine the direction of **F**$_f$. It is a *response* force (Things to Know, item two) and we don't know whether the vector sum of **T**$_2$ and **W**$_\parallel$ (the two other forces that act parallel to the incline, in opposite directions) is directed toward the top or bottom of the incline. In other words, we don't know the direction of the force to which **F**$_f$ must respond in order to keep the mass on the incline at rest.

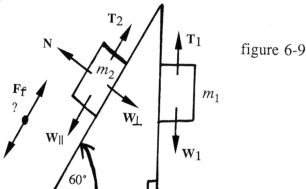

figure 6-9

Two equations can be written regarding this mass, m_2 (the mass at rest on the incline), based on the fact that it is in a state of equilibrium:

$N + W_\perp = 0$ (equilibrium perpendicular to the incline)

$T_2 + W_\parallel + F_f = 0$ (equilibrium parallel to the incline)

Two forces act on the other mass, m_1, hanging at rest to the right of the incline. A force of tension, T_1, exerted by the rope, acts on this mass in the vertically upward direction and its weight, W_1, (equal in magnitude to the weight of m_2) acts on it in the vertically downward direction. Since this mass is also in a state of equilibrium we write:

$T_1 + W_1 = 0$

LINK UNKNOWN TO GIVEN

In terms of the magnitudes of the forces we can proceed as follows:

$T_1 = W_1 = m_1 g = (10)(9.8) = 98\ N = T_2$

$W_\parallel = W_2 \cos 30°$

$W_\parallel = m_2 g \cos 30°$

figure 6-10

$W_\parallel = (10)(9.8)(.866) = 84.87\ N$

$W_\perp = W_2 \sin 30°$

$W_\perp = m_2 g \sin 30°$

$W_\perp = (10)(9.8)(.5) = 49\ N$

$N = W_\perp = 49\ N$

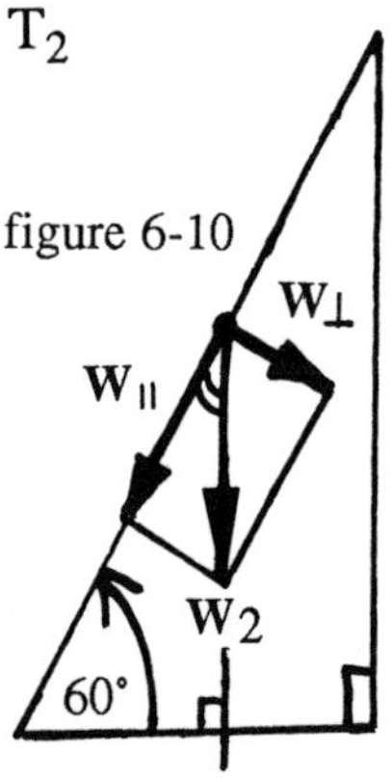

Now we return to the vector equation $T_2 + W_\parallel + F_f = 0$. Since the vectors on the left side of this equation are all directed in the same or in opposite directions, the magnitude of their resultant can be obtained by adding those directed one way (toward the top of the incline), then subtracting those directed the opposite way (toward the bottom of the incline). Thus,

Friction

$98 - 84.87 + F_f = 0$ (magnitude of T_2 = magnitude of T_1)

$F_f = -13.13$ N

The negative value of F_f indicates that the force of friction is directed toward the bottom of the incline. This follows from the fact that we chose to subtract the magnitudes of the forces directed toward the bottom of the incline from those directed toward the top of the incline, rather than the other way around.

We should now verify that F_f is no greater than $\mu_s N$ - otherwise the relationship $F_f \leq \mu_s N$ would be violated and our solution is nonsense.

$F_f \leq \mu_s N$ (magnitude only)
$13.13 \leq (.4)(49)$
$13.13 \leq 19.6$ √   (all is well)

Note: To have substituted $\mu_s N$ in place of F_f in the above equations (as some may be tempted to do in cases such as this) would have produced incorrect values for T, and m_1 would not be in a state of equilibrium. Check it out! The reason such an approach does not work and cannot be used is that the force of static friction need not equal $\mu_s N$, it only cannot exceed it.

6.4   ON YOUR OWN PROBLEMS

(Asterisk indicates solution appears in appendices.)

*(1) How long does it take friction to stop a chair that slides on a horizontal floor if the coefficient of kinetic friction between the chair and the floor is .2 and the chair is moving at 20 m/s?

*(2) The coefficient of static friction between a block and an incline is (.3). The block rests on the incline as the incline is slowly

tilted away from the horizontal (figure 6-11). At what angle from the horizontal will the block begin to slide down the incline?

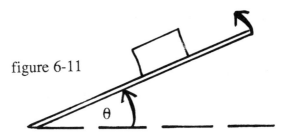

figure 6-11

*(3) A crate that slides on a horizontal floor at the rate of 30 m/s is stopped by friction in 6 meters. What is the coefficient of kinetic friction between the crate and the floor?

*(4) Two blocks connected by a string slide down an incline as illustrated in figure 6-12. The mass of the upper block is 20 kg, that of the lower block is 10 kg. The coefficient of kinetic friction

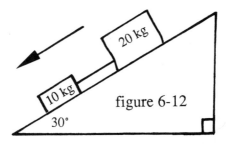

figure 6-12

between either block and the incline is (.2). What is the acceleration rate of the system (both blocks) and the tension in the string, assuming the string is pulled taut between the masses?

(5) A 400 N crate is pulled along a level floor with a rope that makes an angle of 60° with the horizontal. The tension in the rope remains at 200 N as the crate continues to move at the constant rate of 3 m/s. What is the coefficient of kinetic friction between the crate and the floor?

(6) A box slides to a stop on a horizontal floor under the

influence of friction. If the box was moving at the rate of 60 m/s and it took 10 seconds to stop, what is the coefficient of kinetic friction between the box and the floor?

(7) A 10 kg object requires 5 N of force to start moving over a horizontal surface. What is the coefficient of static friction?

(8) If a 15 N force is applied horizontally to the object in problem seven, what is the net force on and acceleration of the object? Assume the coefficients of static and kinetic friction are equal to each other.

(9) How much force must be applied to a 12 N cart to keep it moving at constant speed over a horizontal surface where the coefficient of kinetic friction is .4 ?

(10) If the normal force between two surfaces is tripled, the force of kinetic friction _____.

(11) If the normal force between two surfaces is tripled, the coefficient of static friction _____.

(12) What happens to the magnitude of the force of kinetic friction between a crate and the floor when a weight is placed on top of the crate?

(13) Is the task of pushing a heavy piece of furniture over a horizontal floor made easier or more difficult if the furniture is turned over to a side of smaller surface area? Or, does it not matter? Assume the same surface texture on all sides of the furniture.

(14) A 20 kg box is launched up an incline with a speed of 18 m/s. The coefficient of kinetic friction between the box and the incline is .2 and the angle of inclination is 60°. How far up the incline will the box travel by the time it comes to a stop?

(15) A 30 N package is pulled up an incline at the constant rate of 2 m/s. The angle of inclination is 30° and the force applied to the

package (pulling it parallel to and up the incline) is 20 N strong. How strong would the applied force (still directed parallel to and toward the top of the incline) have to be for the same package to be *lowered* on the same incline at the same constant rate?

(16) A crate weighing 300 N is to be lowered on an incline at a slow and steady pace. The coefficient of kinetic friction between the crate and the incline is .15 and the angle of inclination is 30°. Will the crate need to be held back or pulled down? How much force will need to be applied parallel to the incline?

(17) A magician is about to yank a tablecloth from under a set of fragile dishes. The coefficient of static friction between the dishes and the tablecloth is (.4). How ought the tablecloth be pulled in order that the dishes do not move along with it?

(18) A 30 kg box is to be moved across a level floor. The coefficient of kinetic friction between the box and the floor is (.3). Is it easier to push down on the box at some angle with the horizontal or to pull up on the box at the same angle with the horizontal? Explain your answer, then verify it by determining the force in each case if the angle with the horizontal is 40°.

7

CHAPTER SEVEN

TWO DIMENSIONAL MOTION

7.1 VOCABULARY AND FORMULAS

WORD	SYMBOL	UNIT
Centripetal force	F_c	Newton, N
Period	T	Second, s
Frequency	f	Cycles/second, s^{-1} (also Hertz, Hz)

DEFINITIONS

DIMENSION - Each of three lines that can be arranged to intersect through a common point in a mutually perpendicular manner (figure 7-1).

CENTRIPETAL FORCE - Name for the net (resultant) force acting on an object as it circles around a point at constant speed. This force vector is directed toward the center of the circle.

CENTRIFUGAL FORCE - Used in two ways. (1) Not a force at all, but the *tendency* of a circling object to fly out of the circle, tangentially to the circle. In the absence of the centripetal force this

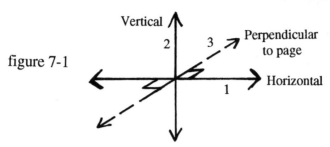

figure 7-1

is indeed what happens. (2) The equal and opposite twin (of the action-reaction pair) of the centripetal force. Used this way, the term *centrifugal force* refers to the force that acts on the object that exerts the centripetal force. The centrifugal force is directed away from the center of the circle.

PERIOD - The time it takes a circling object to complete one revolution.

FREQUENCY - The number of revolutions completed by a circling object per second.

FORMULAS

a. $F_c = mv^2/r$ **b.** $T = 1/f$ and $f = 1/T$ **c.** $F_c = 4m\pi^2 rf^2$
d. $F_c = 4m\pi^2 r/T^2$ **e.** $F_c = m\ a_c$ **f.** $a_c = v^2/r$
g. $a_c = 4\pi^2 rf^2$ **h.** $a_c = 4\pi^2 r/T^2$

7.2 THINGS TO KNOW

(1) Two (or three) dimensional motion is treated as two (or three) independent motions, one in each dimension, occuring simultaneously. The laws and formulas of physics are applied separately to the motion in each dimension.

(2) Ask yourself: What type of motion and what conditions exist

Two Dimensional Motion 159

in the horizontal dimension and what type of motion and what conditions exist in the vertical dimension (or any other set of perpendicular dimensions). Then apply the appropriate principles and formulas to each dimension, one dimension at a time.

(3) Projectile motion problems on earth are best divided into horizontal and vertical motions, with gravity acting only in the vertical dimension. To avoid confusion attach h and v subscripts to the symbols that appear in the equations. For example, the equation $d = vt$ should be written as $d_H = v_H t$, thereby making sure that v will be the horizontal velocity, not the vertical or resultant velocity. Analogously, the equation $v_f = v_i + at$ applied to the vertical dimension, should be written as $v_{fv} = v_{iv} + at$.

(4) Care must be taken when assigning v-initial values in each dimension. An object launched horizontally, for example, at a rate of 10 m/s has an initial velocity in the horizontal dimension of 10 m/s and an initial velocity of *zero* m/s in the vertical dimension. An object launched at an angle of 30° from the horizontal with a velocity of 100 m/s has an initial velocity in the horizontal dimension of 87 m/s (the component of the initial velocity vector in the horizontal dimension) and an initial velocity in the vertical dimension of 50 m/s (the component of the initial velocity vector in the vertical dimension). The initial velocity in either dimension is *not* 100 m/s. The 100 m/s velocity vector at 30° is the resultant of the initial velocities in the two dimensions.

(5) An object circling around a point at constant speed is *not* in a state of equilibrium, despite the fact that its speed is not changing. A net force must be acting on it since its direction of motion *is* changing. This force is called the *centripetal force*, F_c. But, you may wonder, is not $F = ma$ and, at constant speed, is not *a* equal to

zero? No! Circular motion is two dimensional motion and the laws of physics, including $F_N = ma$, are not supposed to be applied to the resultant motion but to the independent motion that occurs in each dimension. As an object circles around a point, its velocity in each dimension does change. There *is* therefore an acceleration and a force in each dimension. It is just that in uniform circular motion things are so rigged that the magnitude of the object's resultant velocity (its speed) does not change. But this fact, while interesting, has no bearing on the applicability of the laws of motion (except those formulas that are specifically designed to be based on resultant quantities, such as $F_c = mv^2/r$, in which the symbol v represents the magnitude of the resultant velocity, and formulas (f), (g) and (h), in which a refers to the magnitude of the resultant acceleration rate).

(6) Centripetal force does not lead an independent existence. It is merely the name given to the resultant force that acts on a circling object. That net force must have a physical basis for existing - gravity, friction, tension, and so on. *Centripetal force* does not provide such a basis. When adding the forces that act on a circling object to find their resultant, we cannot include "centripetal force" in the list. Instead, the centripetal force *is* the resultant of all the forces present, each of which has a physical basis for existing. Similarly, when we seek to determine why an object is circling, we cannot attribute the motion to "centripetal force". We must find forces that are physically real.

(7) Be careful about the difference between frequency and period - they are easily confused. Period is seconds per cycle, frequency is cycles per second. An object that "completes a round trip once every eight seconds" has a period of 8 seconds and a frequency of .125 cycles/second (Formulas, item *b*).

7.3 SOLVING THE PROBLEMS

(1) A stone is launched horizontally with a speed of 20 m/s from the top of a cliff 100 meters high. (a) How long does it take the stone to reach the ground below? (b) How far from the base of the cliff does the stone strike the ground? (c) At what angle does the stone strike the ground? (d) What is the speed of the stone as it impacts into the ground? Ignore air resistance.

SKETCH AND IDENTIFY

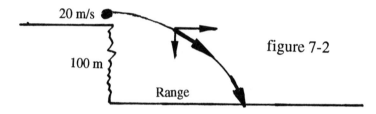

figure 7-2

This is a two dimensional motion problem. There is activity in the horizontal dimension because the stone was launched horizontally, and there is activity in the vertical dimension because gravity acts in that dimension. The following facts should be kept in mind regarding the motion in each dimension:

HORIZONTAL	**VERTICAL**
No force	Gravity is net force
Constant velocity	Accelerated motion, free fall
$v_H = 20$ m/s	Initial velocity is zero, $v_{iv} = 0$
$d_H = v_H t$	$v_{fv} = v_{iv} + at$, $a = 9.8$ m/s^2
	$d_v = v_{iv} t + at^2/2$

If there were no force of gravity, the stone would continue to move horizontally forever and never reach the ground. The only reason the stone gets closer to the ground is because there exists the vertically directed pull of gravity. So in seeking the time it takes to reach the ground (part a) we need concern ourselves only with the stone's vertical motion. No matter what the stone's horizontal speed, it will reach the ground in the same time (assuming it is launched from the same height). The stone's horizontal speed makes no contribution to its getting closer to the ground!

Once the stone strikes the ground all motion comes to an end (at least that is the implication in the problem). That puts a limit on how long the horizontal motion lasts. In other words, the horizontal motion comes to an abrupt end at a time dictated by developments in the vertical dimension.

LINK UNKNOWN TO GIVEN

(a) The time to reach the ground is decided by activity in the vertical dimension:

$d_v = v_{iv}t + at^2/2$ (vertical, downward trip)

$100 = (0)t + (9.8)t^2/2$

$t = 4.52$ sec

(b) The range is decided by activity in the horizontal dimension:

$d_H = v_H t$ (horizontal trip)

$d_H = (20)(4.52) = 90.4$ meters

(c) As time marches on, the horizontal component of the stone's velocity remains fixed at 20 m/s (to the right) but the vertical component (downward) increases at the rate of 9.8 m/s^2. This causes the resultant velocity vector of the stone to take on an increasingly vertical tilt.

Two Dimensional Motion

After 4.52 seconds, as the stone strikes the ground, its horizontal velocity is 20 m/s to the right and its vertical velocity downward can be determined as follows:

$v_{fv} = v_{iv} + at$ (vertical, downward trip)

$v_{fv} = 0 + (9.8)(4.52)$

$v_{fv} = 44.3$ m/s

The stone's velocity as it strikes the ground consists therefore of two components, as illustrated in figure 7-3.

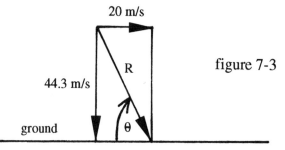

figure 7-3

The angle at which the stone strikes the ground can be determined (from figure 7-3) as follows:

$\tan \theta = 44.3/20$

$\theta = 66°$

(d) The magnitude of the stone's resultant velocity as it strikes the ground can be determined with the aid of the pythagorean theorem:

$R^2 = 44.3^2 + 20^2$

$R^2 = 48.6$ m/s

(2) A baseball is struck by a bat and launched at an angle of 53° from the horizontal with a speed of 30 m/s. (a) How high does the baseball rise? (b) How far does it travel horizontally before returning to the ground? Assume the baseball was launched at ground level and ignore air resistance.

SKETCH AND IDENTIFY

As the baseball is launched it is provided with horizontal and vertical motion. Its initial horizontal velocity is the horizontal component of its velocity vector at the time of launch, and its initial vertical velocity is the vertical component of the same vector. These are calculated in figure 7-4.

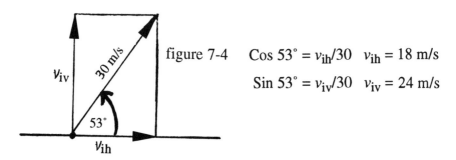

figure 7-4 $\cos 53° = v_{ih}/30$ $v_{ih} = 18$ m/s
              $\sin 53° = v_{iv}/30$ $v_{iv} = 24$ m/s

As time goes on, the horizontal component of the ball's velocity remains fixed at 18 m/s. The vertical component, on the other hand, decreases on the way up at the rate of 9.8 m/s² until it becomes equal to zero (when the ball stops rising and is at its highest point), then reverses direction and increases (on the way down) at the same rate. At the highest point in its path the ball moves horizontally only - its vertical velocity having been reduced to nothing (figure 7-5).

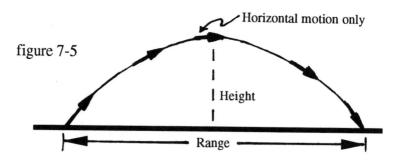

figure 7-5

The horizontal and vertical motions have the following characteristics:

Two Dimensional Motion

HORIZONTAL	**VERTICAL**
No force	Gravity is net force, free fall
Constant velocity	Accelerated motion, $a = -9.8$ m/s^2
$v_H = 18$ m/s	$v_{iv} = 24$ m/s, $v_{fv} = v_{iv} + at$
$d_H = v_H t$	$d_v = v_{iv}t + at^2/2$

Note: The vertical distance is measured from the ground, with the ground serving as *reference point* (recall chapter three). This makes the velocity positive on the way up (going away from the reference point) and negative on the way down (going toward the reference point). Since the magnitude of the ball's velocity decreases on the way up and increases on the way down, the acceleration is negative both ways (section 3.1, graphs H and L).

LINK UNKNOWN TO GIVEN

(a) The height is determined by activity in the vertical dimension only, as such:

$v_{fv} = v_{iv} + at$ (vertical, upward trip)

$0 = 24 - 9.8t$ (vertical velocity at highest point is zero)

$t = 2.45$ sec

The ball rises for 2.45 seconds, falls for another 2.45 seconds, and returns to the ground after 4.9 seconds.

$d_v = v_{iv}t + at^2/2$ (vertical, upward trip)

$d_v = (24)(2.45) + (-9.8)(2.45)^2/2$

$d_v = 29.4$ meters   (the height)

(b) The range (horizontal distance) is determined by activity in the horizontal dimension only:

$d_H = v_H t$ (entire horizontal trip)

$d_H = (18)(4.9)$

$d_H = 88.2$ meters

A little arithmetic also reveals that the ball strikes the ground in the manner that it was lunched - at 53° from the horizontal, with a speed of 30 m/s.

(3) A 3 kg mass is circling at constant speed in a horizontal plane at the end of a string. The string is 1.5 meters long. If the object makes five revolutions every second, (a) how strong is the centripetal force acting on it, and (b) what is the centripetal acceleration rate? (c) If the maximum tension the string can withstand is 400 N, how fast can the object rotate without the string snapping?

SKETCH AND IDENTIFY

A horizontal plane is one that is parallel to the ground. The centripetal force in such a circle must be directed parallel to the ground, so gravity cannot be contributing to it. The centripetal force in this problem can therefore be provided only by the tension in the string. The object at the end of the string tries to fly out tangentially to the circle, due to its own inertia. But that can only happen if the string is stretched (the other end of the string is held fixed at the circle's center) and the string strenuously resists being stretched. The string pulls the object inward with just enough force to prevent its own length from increasing. This keeps the object on its circular course, with the radius of the circle equal to the unchanging length of the string, 1.5 meters in this case.

Since we are provided with the frequency of rotation, let us use the frequency formula for centripetal force.

LINK UNKNOWN TO GIVEN

(a) $F_c = 4m\pi^2 r f^2$

Two Dimensional Motion

$F_c = 4(3)(3.14)^2(1.5)(5)^2$

$F_c = 4441.3$ N

(b) Once F_c is known it is a straight forward matter to go to $F_c = ma_c$ to find the centripetal acceleration:

$F_c = ma_c$ (magnitude only)

$4441.3 = (3)a_c$

$a_c = 1480.4$ m/s²

To find a_c without making use of F_c, such as when the mass is not given, proceed as follows:

$a_c = 4\pi^2 rf^2$

$a_c = (4)(3.14)^2(1.5)(5)^2$

$a_c = 1480.4$ m/s²

(c) The faster a given mass rotates around a given point, the greater the centripetal force needed to keep it in its circular track. Since the centripetal force in this problem is provided by the tension in the string and the maximum tension available is 400 N, the maximum possible speed is determined as follows:

$mv^2/r \leq 400$

$3v^2/1.5 \leq 400$

$v^2 \leq (400)(1.5)/3$

$v \leq 14.14$ m/s

(4) In the apparatus illustrated in figure 7-6 the vertical pole is rotating around an axis parallel to and through itself. The cord is 2 meters long and the horizontal bar is .5 meters long. What should be the period of rotation in order that the cord makes an angle of 30° with the vertical?

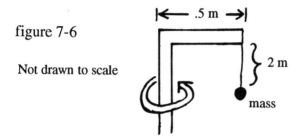

figure 7-6

Not drawn to scale

SKETCH AND IDENTIFY

When the cord is tilted 30° from the vertical the scene appears as illustrated in figure 7-7. There are then two forces acting on the rotating mass: tension and gravity. These force vectors, **T** and **W**, are arranged as illustrated in figure 7-8. Since the mass circles in a

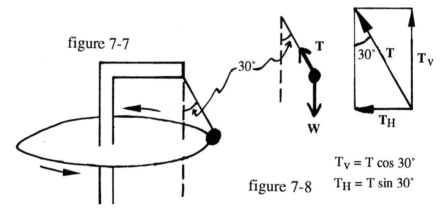

figure 7-7

figure 7-8

$T_V = T \cos 30°$
$T_H = T \sin 30°$

horizontal plane, it must be in equilibrium in the vertical dimension. This implies that $W = T_V$ (magnitude only, no vector symbol). Since **W** has no horizontal component, the centripetal force is provided only by the horizontal component of the tension. Thus, it must be true that $\mathbf{F_c} = \mathbf{T_H}$.

Now, the radius of the circle is equal to the sum of the length of the horizontal bar and the leg of the right triangle formed by the cord that is parallel to the plane of the circle (side X in figure 7-9). This makes the radius 1.5 meters long.

Two Dimensional Motion

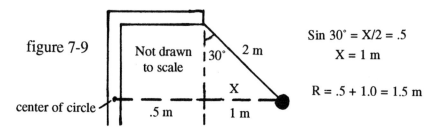

figure 7-9

Sin 30° = X/2 = .5
X = 1 m
R = .5 + 1.0 = 1.5 m

LINK UNKNOWN TO GIVEN

$W = T_v$ (magnitude only)

$mg = T \cos 30°$ (see figure 7-8)

$T = mg/\cos 30°$

$T_H = T \sin 30° = (mg/\cos 30°)(\sin 30°)$

$T_H = mg \tan 30°$

$F_c = T_H$ (magnitude only)

$4m\pi^2 r/T^2 = mg \tan 30°$

The symbol *T* here - in *italics* - represents the period of rotation, not the tension. The masses drop out, as expected, and we get:

$T^2 = (4)(3.14)^2(1.5)/(9.8)(.577)$

$T = 10.47$ seconds

(5) An anti-aircraft gun makes an angle of 40° with the horizontal and its muzzle velocity is 2000 m/s. For what point(s) in time after launch should the fuse in a shell (launched by the gun) be set in order that the shell explodes at an altitude of 3000 meters?

SKETCH AND IDENTIFY

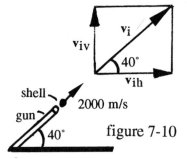

figure 7-10

$v_{ih} = (2000)(\cos 40°) = 1532.1$ m/s

$v_{iv} = (2000)(\sin 40°) = 1285.6$ m/s

The time it takes the shell to reach an altitude of 3000 meters is determined entirely by activity in the vertical dimension. The shell may be at that altitude at two points in its flight, as illustrated in figure 7-11.

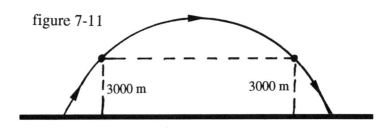

figure 7-11

LINK UNKNOWN TO GIVEN

$d_v = v_{iv}t + at^2/2$

$3000 = (1285.6)t + (-9.8)t^2/2$

This type of quadratic equation can, in principle, have either two solutions for t (as in figure 7-11), one solution (in which case 3000 meters is the highest point), or no solutions (in which case the highest point is lower than 3000 meters). To determine the correct scenario for our problem and to solve for t we first put the equation into *standard form*: $ax^2 + bx + c = 0$.

$4.9t^2 - 1285.6t + 3000 = 0$

Next, we simplify and divide every term by 4.9. That yields:

$t^2 - 262.4t + 612.2 = 0$

Employing the standard solution form for such equations yields:

$x = (-b \pm \sqrt{b^2 - 4ac})/2a$

$t = (262.4 \pm \sqrt{68853.8 - 4 \cdot 1 \cdot 612.2})/2 \cdot 1$

Two solutions emerge: $t_1 = 2.4$ seconds and $t_2 = 260$ seconds. Either time can be employed to set the fuse in order that the shell explodes at an altitude of 3000 meters.

Two Dimensional Motion　　　　　　　　　　　　　　　　　　　171

(6) A 4000 N automobile rounds a curve of radius 600 meters on an unbanked road. The speed of the automobile remains constant 80 km/hr. (a) What must be the minimum coefficient of static friction between the tires and the road in order that the automobile does not skid? (b) At what angle should the roadbed be banked so that an automobile of the same weight, moving at the same speed around the same curve, does not skid even when the road is covered with ice and friction is reduced to negligibility?

SKETCH AND IDENTIFY

When a car rounds a curved road it goes through an arc of a circle (or, at least, a close approximation to it) whose plane is parallel to the ground (a *horizontal* circle). A centripetal force must be present for the car to stay on its curved course, otherwise it slides out of the circle - a condition referred to as a *skid*. The centripetal force must be directed parallel to the ground, toward the center of the circle.

On a level road this centripetal force is provided by friction between the tires and the ground. Friction compels the car to follow in the direction in which the wheels point so that the tires roll along on the ground, instead of sliding and scraping over it. When the direction of the wheels is changed, friction forces the direction of motion of the car to change along with it (figure 7-12). Since this force of friction acts to prevent sliding (of tires over road) *before* such sliding occurs, it is appropriately placed in the catagory of *static* friction.

On a banked road the burden of providing the centripetal force is shared by friction and the component of the normal force that acts parallel to the ground (and therefore also parallel to the plane of the

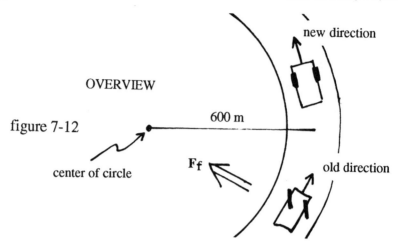

figure 7-12

circle). If there is no friction (such as when the road is covered with ice) the burden shifts entirely to the normal force. The force of gravity can make no contribution to the centripetal force since it is directed perpendicularly to the plane of the circle. No component of gravity acts parallel to the plane of the circle (figure 7-13).

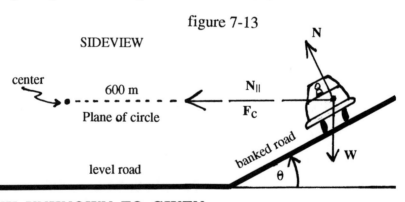

figure 7-13

LINK UNKNOWN TO GIVEN

(a) On a level road:

$$F_c = F_{NET} = F_f + N + W$$

Since equilibrium exists in the vertical dimension:

$N + W = 0$ (figure 7-14)

$F_c = F_f$ (magnitude only)

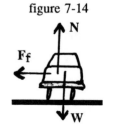

figure 7-14

Two Dimensional Motion

$$mv^2/r = F_f \leq \mu_s N \quad \text{(chapter six)}$$

$$N = W = mg \quad \text{(magnitude only)}$$

$$mv^2/r \leq \mu_s mg$$

$$v^2/r \leq \mu_s g$$

$$\mu_s \geq (22.2)^2/(9.8 \cdot 600) \quad (80 \text{ km/hr} = 22.2 \text{ m/s})$$

$$\mu_s \geq .08$$

(b) No friction, banked road:

$$\mathbf{F_c} = \mathbf{F_{NET}} = \mathbf{N} + \mathbf{W}$$

$$F_c = mv^2/r = (408.2)(22.2)^2/600 = 335 \text{ N} \quad (m = W/g)$$

$W/N = \cos\theta$ (fig. 7-15)

$4000/N = \cos\theta$ (equation 1)

$F_c/N = \sin\theta$ (fig. 7-15)

$335/N = \sin\theta$ (equation 2)

Rewriting equations 1 and 2 yields:

$4000/\cos\theta = N$ and $N = 335/\sin\theta$

$4000/\cos\theta = 335/\sin\theta$

$\sin\theta/\cos\theta = 335/4000$

$\tan\theta = .084$

$\theta = 4.8°$ (angle of embankment)

figure 7-15

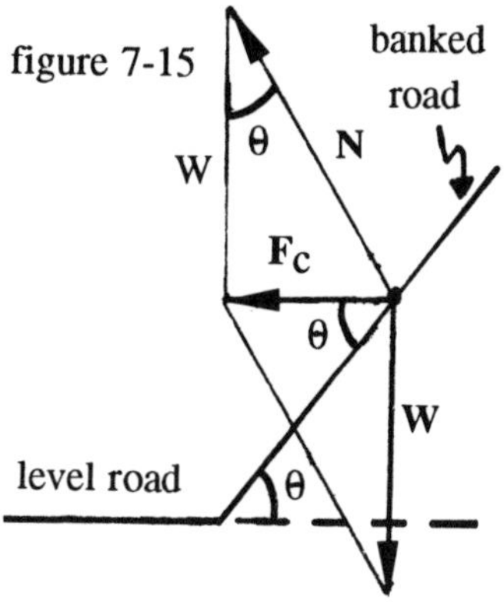

Note: In this problem, N cannot be made equal to the component of W perpendicular to the banked roadway, as we did previously in other problems, because a state of equilibrium does *not* exist in the dimension perpendicular to the banked roadway. Indeed, the net force has a component perpendicular to the roadway, guaranteeing that equilibrium does not exist in that dimension.

7.4 ON YOUR OWN PROBLEMS

(Asterisk indicates solution appears in appendices.)

*(1) A diver is to be launched horizontally off a cliff that is 80

meters high. What minimum launch velocity enables her to clear the rocks below, if those rocks extend outward from the base of the cliff a distance of 30 meters (figure 7-16).

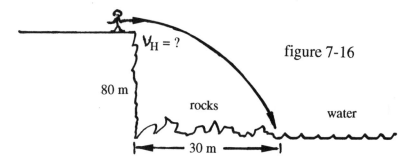

*(2) A projectile is launched with an initial velocity of 200 m/s. What must be the launching angle (from the horizontal) if the projectile is to strike a target 500 meters away?

*(3) The radius of the moon's orbit around the earth is 3.6×10^8 meters. The moon's period of rotation is 27.3 days. What is the acceleration rate of the moon toward the earth?

*(4) A phonograph turntable rotates at the constant rate of 33 rev/minute. A small object placed on the turntable remains in place on the turntable if it is situated within 4 cm from the center. If placed farther from the center, the object slides on the turntable. What is the coefficient of static friction between the object and the turntable?

*(5) A bomber moving at the rate of 150 m/s is to release its load while diving at an angle of 40° from the horizontal. The bombs are to be released at an altitude of 10,000 meters. How far from the target (measured along the ground) should the bombs be released in order that they strike the target? Assume no air resistance.

*(6) A train rounds a curve at the constant rate of 20 m/s. An object hangs at the end of a rope that is suspended from the ceiling inside the train. The radius of curvature of the tracks is 300 meters.

Two Dimensional Motion   175

What angle will the rope make with the vertical as the train rounds the curve?

(7) A rock is launched horizontally from the roof of a building that is 2000 meters tall. (a) How long does it take the rock to reach the ground below? (b) How far from the building does the rock strike the ground if it was launched at 30 m/s?

(8) How fast is the rock in problem seven moving as it collides with the ground? At what angle does it strike the ground?

(9) A projectile is launched at an angle of 40° from the horizontal. With what speed must the projectile be launched in order that it land at a point 1500 meters distant from the launching point? Neglect air resistance.

(10) A rescue plane is to drop a rubber raft to the survivors of a shipwreck. How far from the site of the shipwreck (measured along the water) should the airplane be when it releases the raft if the plane is moving horizontally at the rate of 60 m/s, at an altitude of 4000 meters?

(11) A person standing on earth at the equator is actually revolving around the center of the earth, completing one revolution every 23 hours and 56 minutes. The radius of the giant circle he goes through every day is 6400 km. (a) How strong a centripetal force is required to keep a 100 kg person circling in this manner? (b) Is the pull of gravity (the person's weight) sufficient to provide this force?

(12) A 1000 kg car is to round a curve of radius 200 meters on an unbanked road. The coefficient of static friction between the tires and the road is (.4). How fast can the car be traveling as it rounds the curve without getting into "a skid" (no slipping or sliding over the surface of the road)?

(13) Would the answer to problem twelve be different if the car

were heavier? In general, does driving a heavier car provide more protection against skidding, assuming all other pertinent factors (such as the composition of the road and tires) are kept the same?

(14) A hunter aims her rifle directly at a target 300 meters away. The bullet is launched horizontally with a speed of 450 m/s. By how much does the bullet miss the target? Assume no air resistance.

(15) A baseball player can throw a ball a maximum distance of 100 meters, measured along the ground. How high can he throw the ball? (Hint: determine how much speed he can impart to the ball.)

(16) A projectile is launched with an initial speed of 50 m/s, at an angle of 60° from the horizontal. Determine the vertical and horizontal position of the projectile three seconds after it is launched.

(17) What are the horizontal and vertical components of the velocity of the projectile in problem sixteen, at the time indicated in that problem?

(18) What must be the period of rotation of the earth in order that a person standing on a scale at the equator reads *zero* Newtons?

(19) A player kicks a football at an angle of 30° from the horizontal, launching the ball with a speed of 20 m/s. A second player standing 400 meters away starts running to catch the ball at the instant it is kicked. How fast should she run in order to catch the ball before it hits the ground?

(20) An 8 kg object at the end of a rope is made to rotate through a vertical circle (plane of circle is perpendicular to the ground). The radius of the circle is 2 meters and the speed of rotation is kept constant at 6 m/s. (a) What is the tension in the rope when the object is at the top of the circle? (b) What is the tension when the object is at the bottom of the circle? (c) At what point in such a circle is the rope most likely to snap?

Two Dimensional Motion

(21) In an exciting carnival ride, passengers are placed against the inside of a circular wall that forms the perimeter of a round, rotating platform. When the platform attains a high enough frequency of rotation the floor is pulled out from under the passengers and, if all goes well, the passengers do not fall through the bottom of the rotating vehicle. The radius of one such platform is 4 meters and the coefficient of static friction between the passengers and the wall they lean against is (.3). With what minimum frequency must the vehicle be rotating in order that the passengers do not fall when the floor is pulled out from under them?

(22) A 5 kg stone attached to a string is whirled in a horizontal circle of radius 2 meters. The string dips down from the horizontal at an angle of 40°. What is the tension in the string and the speed of the stone?

(23) A 40 kg pilot comes out of a 90 m/s vertical dive in a circular arc such that her upward acceleration rate is 6g. What force is exerted on her by the airplane seat (to which she is bound) at the bottom of the arc, assuming the airplane's speed remains constant?

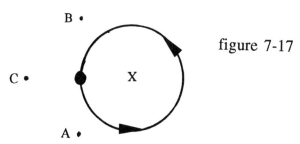

figure 7-17

(24) The object in figure 7-17 is rotating at a uniform rate around point X, in a counter-clockwise manner. In the position shown, the object's velocity is directed toward point _____ , its centripetal acceleration is directed toward point _____ and the centripetal force is directed toward point _____ .

8

CHAPTER EIGHT

GRAVITY

8.1 VOCABULARY AND FORMULAS

DEFINITIONS

UNIVERSAL GRAVITATION - Force exerted by every mass in the universe on every other mass. The mechanism by which this force is exerted is through the action of the *gravitational field.*

GRAVITATIONAL FIELD - Entity permeating the space in the vicinity of a mass. This entity acts to exert a force on any other mass that enters the field.

ORBITAL VELOCITY (V_o) - Velocity required to maintain a stable circular orbit when the force of gravity provides the centripetal force.

ESCAPE VELOCITY (V_e) - Minimum velocity required at point of launch in order that the launched object never returns to its starting point under the influence of the gravitational force pulling it backward, toward the body from which the object is escaping.

FORMULAS

a. $F = Gm_1m_2/d^2$ $G = 6.67 \times 10^{-11}$ N·m²/kg²   **b.** $V_o = \sqrt{GM/R}$

Gravity

c. $V_e = \sqrt{2GM/R}$ d. $V_o = \sqrt{gr}$ e. $T^2/a^3 =$ constant $= k$
f. $F \propto m_1 m_2 / d^2$ g. $R_e = 6400$ km $= 6.4 \times 10^6$ m
h. $M_e = 6 \times 10^{24}$ kg

8.2 THINGS TO KNOW

(1) The formula $F = Gm_1m_2/d^2$ is designed for *point masses* - objects whose dimensions are small compared to the distance between them. If this is not the case the formula cannot be used. Spherically symmetric masses, however, can be treated as point masses even if the spheres are large compared to the distance between them. When this is done, each sphere's entire mass is treated as if it is concentrated at the center of the sphere and d in the formula, consequently, represents the distance between the centers.

(2) The distance that counts in the case of an object being influenced by the earth's gravitational pull is the distance between the center of the object and the center of the earth. At sea level this quantity is, for all practical purposes, equal to the radius of the earth, R_e (Formulas, item *g*). Under no circumstances can the distance between the object and the ground be plugged into d in the formula for gravity.

(3) The formula $F = Gm_1m_2/d^2$ provides the force exerted on either of two masses by and toward the other. These forces are equal in magnitude and opposite in direction, as dictated by the principle of action-reaction, even if the masses are not equal (figure 8-1). The force exerted upward on the earth by a falling apple is equal to the force exerted downward on the apple by the whole earth.

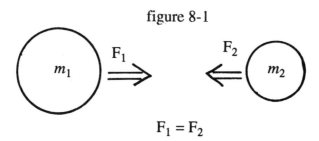

figure 8-1

$F_1 = F_2$

(4) Objects can be said to be "falling freely" and "accelerating toward earth" even when they are not moving toward the earth. Objects that decelerate on the way up and satellites in orbit are prime examples of this. Both are "falling" and "accelerating toward" the earth in the sense that their **F** and **a** vectors are directed toward the earth (despite the fact that their **v** vectors are not so directed).

(5) In the formula $V_o = \sqrt{GM/R}$, the symbol M represents the mass of the body at the center of the circular orbit (not that of the object doing the orbiting) and R represents the radius of the circle (not the radius of either body).

In the formula $V_e = \sqrt{2GM/R}$, the symbol M represents the mass of the body from which an object is escaping (not that of the object doing the escaping) and R represents the distance between the point of launch and the center of the body represented by M.

(6) The formula $V_o = \sqrt{gr}$ is designed to serve as a good approximation for objects orbiting near the surface of the earth, such as within 100 km from the surface. The symbol r represents the radius of the earth and g is 9.8 m/s².

(7) In the formula $T^2/a^3 = k$ (Kepler's third law), the symbol a represents the semi-major axis of the elliptical orbit of a planet around the sun. If the orbit is nearly circular (as is the case for the five planets closest to the sun), a becomes the radius of the orbit.

Gravity

The formula can be applied to any group of bodies that orbit around one body - the nine planets around the sun, Jupiter's sixteen moons around Jupiter, the many man-made satellites that together with the moon orbit around our earth, and so on. For each such group, however, there is a unique value for k. In the case of the nine planets around the sun, k is 2.985×10^{-19} sec^2/m^3; for the satellites that orbit around the earth k is 9.8×10^{-14} sec^2/m^3. All the members of a group have the same ratio of T^2/a^3, equal to the value of k for that group.

(8) The force of gravity acting on an object is the object's weight. But, whereas the formula $F = Gm_1m_2/d^2$ is applicable universally, for all masses and distances, the formula $W = mg$ (with $g = 9.8$ m/s^2) is applicable only to objects on earth, at sea level.

(9) The formula $V_o = \sqrt{GM/R}$ is applicable only to stable, circular orbits. It cannot be used for an orbiting object that spirals away from or toward the body that is acting on it gravitationally or when an orbit is changing to a larger or smaller one.

(10) Objects are *not* weightless in space. Indeed it is the force of gravity (in other words, the object's weight) that provides the centripetal force that maintains an orbit. For orbits near the surface of the earth, this force is just about equal to the weight of the (orbiting) object on earth! Things seem weightless inside a spaceship because all objects in the spaceship, including the ship itself, accelerate ("fall") toward the earth at the same rate. The situation is very much akin to that of a person in a freely falling elevator - there *seems* to be no gravity inside the elevator because all objects in the elevator fall together with the elevator, at the same rate, toward the earth.

8.3 SOLVING THE PROBLEMS

(1) A 10,000 kg spacecraft is traveling away from Earth. (a) What is its weight at a distance of 12,800 km from the earth's surface? (b) What is its weight at a distance of 14,000 km from the earth's surface?

SKETCH AND IDENTIFY

The only distance that matters, insofar as gravity and weight are concerned, is that between the center of the spacecraft and the *center* of the earth. The 12,800 km distance provided in the problem cannot be plugged into d in the formula for universal gravitation because it's the wrong distance - that between the spacecraft (which we rightfully treat as a point mass) and the *surface* of the earth. That distance is useful only to the extent that from it can be derived the distance that does count. By adding the radius of the earth (6400 km) to the distance between the spacecraft and the surface of the earth (12,800 km) we obtain the distance between the spacecraft and the center of the earth (figure 8-2). That distance is 19,200 km in the case of part a and 20,400 km in the case of part b.

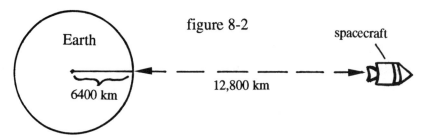

figure 8-2

A "brute force" method for solving this problem consists of plugging all the data into the formula $F = Gm_1m_2/d^2$ and calculating the unknown force, F. We notice, however, that the value for d in

Gravity

part (a) - 19,200 km - is precisely three times the value for d when the spacecraft is on the earth's surface - 6400 km. This enables us to solve part a via the easier method of using the proportionality relationship $F \propto m_1 m_2 / d^2$ (Formulas, item F).

LINK UNKNOWN TO GIVEN

(a) Let us compare the force, F, at 19,200 km to the force at 6400 km.

$$F \propto m_1 m_2 / d^2$$

Since the product $m_1 m_2$ is the same at both distances (same earth and same spacecraft) we may write:

$$F \propto 1/d^2$$

Since d is three times as great at 19,200 km than at 6400 km, d^2 is nine times as great, and $1/d^2$ is one-ninth as great. Since F and $1/d^2$ grow and shrink in tandem (they are proportional to each other) this implies that F, the force of gravity and the spacecraft's weight, is one-ninth as strong at the 19,200 km mark as at the 6400 km mark.

To find the spacecraft's weight at 6400 km from the earth's center we can make use of $W = mg$, since at that distance the spacecraft is on the earth's surface (g is 9.8 m/s² only at sea level).

$W = mg = (10,000)(9.8) = 98,000$ N

The weight at 19,200 km is one-ninth of 98,000 N or 10,889 N.

To see how the "brute force" method works, we solve part b below via that method.

(b) $F = G m_1 m_2 / d^2$

$$F = \frac{(6.67 \times 10^{-11})(6 \times 10^{24})(1 \times 10^4)}{(20.4 \times 10^6)^2} = \frac{(6.67)(6) \times 10^5}{(20.4)^2}$$

$F = 9.6 \times 10^3$ N

(2) A spacecraft is to orbit the earth midway between the earth and the orbit of the moon. Assuming the moon's gravitational pull can be ignored, (a) what must be the spacecraft's velocity if its orbit is to be stable, circular and centered on the earth, (b) what must be the spacecraft's period? (The moon orbits the earth at an average distance of about 384,000 km from the center of the earth - roughly 60 times the radius of the earth.)

SKETCH AND IDENTIFY

The most direct method for determining the spacecraft's velocity is to use the formula for orbital velocity, $V_o = \sqrt{GM/R}$. Now, G and M (the mass of the earth) are known, and R is to be one-half of 384,000 km or 1.92×10^5 km. The formula $V_o = \sqrt{gr}$ cannot be used here because the orbit is not near the earth's surface (Things to Know, item 6).

The most direct method for determining the period of the spacecraft is to use Kepler's third law. Both the spacecraft and the moon orbit around the same body, the earth, so the ratio T^2/a^3 should be the same for both.

LINK UNKNOWN TO GIVEN

(a) $V_o = \sqrt{GM/R}$

$V_o = \left[\dfrac{(6.67 \times 10^{-11})(6 \times 10^{24})}{1.92 \times 10^8} \right]^{1/2}$ (converting km to meters)

$V_o = (20.84 \times 10^5)^{1/2} = (208.4 \times 10^4)^{1/2} = 14.44 \times 10^2$ m/s

$V_o = 1.44 \times 10^3$ m/s

(b) We assume that the orbit of the spacecraft and that of the moon around the earth are both circular. Then a in Kepler's third law can be replaced by R, the radius of either orbit.

Gravity

$T^2/a^3 = T^2/R^3 = $ constant $= k$

The problem does not provide the value of k for bodies orbiting around the earth, but the fact that all such bodies have the same T^2/R^3 ratio enables us to proceed as follows:

$$\frac{T_s^2}{R_s^3} = \frac{T_m^2}{R_m^3} \qquad \text{(subscript } m \text{ for moon, } s \text{ for spacecraft)}$$

$$\frac{R_m^3}{R_s^3} = \frac{T_m^2}{T_s^2} \qquad (R\text{'s refer to radii of orbits)}$$

$$\left|\frac{R_m}{R_s}\right|^3 = \frac{T_m^2}{T_s^2}$$

$$(2)^3 = \frac{T_m^2}{T_s^2} \qquad (R_m \text{ is twice } R_s)$$

$$T_s^2 = T_m^2/8$$

$$T_s^2 = (27.3)^2/8 = 93.16 \text{ days} \qquad (T_m = 27.3 \text{ days})$$

$$T_s = \sqrt{93.16} = 9.65 \text{ days} = 8.34 \times 10^5 \text{ sec}$$

(3) What is the acceleration rate of any freely falling object on the moon? (The mass of the moon is about one-eightieth (1/80) that of the earth and its diameter is about one-fourth (1/4) the diameter of the earth.)

SKETCH AND IDENTIFY

When an object of mass m falls freely near the surface of a body of mass M, it is accelerated by the pull of gravity exerted on it by M. The strength of that pull is provided by the formula:

$$F = Gm_1m_2/d^2 = GMm/R^2$$

where R is the radius of M - the distance between the center of M and the center of m (figure 8-3).

figure 8-3

Now, the formula $F = ma$ is also applicable to the freely falling object, just as it is when any force acts on any mass. So we can write:

$F = ma = GMm/R^2$

The m's drop out and we are left with:

$a = GM/R^2$

On earth, the free fall acceleration rate is 9.8 m/s² because GM_e/R_e^2 is (9.8). On the moon, the free fall acceleration rate may be some other value, since M and R are not the same as on earth. By comparing the value of GM/R^2 on the moon to that on earth we can compare the value of a (the free fall acceleration) on the moon to that on earth.

LINK UNKNOWN TO GIVEN

$a_e = GM_e/R_e^2 = g = 9.8$ m/s²

$a_m = GM_m/R_m^2$ (subscript m for *moon*)

$a_m = G(M_e/80)/(R_e/4)^2 = (16/80)GM_e/R_e^2$

$a_m = (.2)GM_e/R_e^2 = (.2)g = 1.96$ m/s²

(4) A woman stands on a scale at the equator and finds that the scale reads 200 N. Is that her true weight, the pull of gravity on her? If not, what is her true weight?

Gravity 187

SKETCH AND IDENTIFY

Two forces act on a person standing anywhere on earth. There is the pull of the earth's gravity, **W**, directed toward the center of the earth, and there is the normal force, **N**, directed perpendicularly to the surface of the earth, away from the earth's center. While these forces are opposite in direction, they cannot be equal in magnitude. If they were, the resultant force on the person would be zero

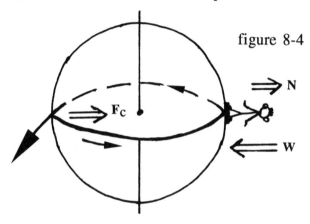

figure 8-4

and there would not exist a centripetal force to keep the person circling around the earth's axis. (Recall - from chapter seven - that the centripetal force is the *net* force acting on an object in uniform circular motion.) He or she would then go flying out into space, tangentially to the surface of the earth. Instead, **W** must be greater than **N** in order that the difference between these forces, directed toward the center of the earth in the case of a person standing at the equator, can provide the net force necessary to prevent such flying out (figure 8-4).

Now, a scale reads the force exerted *on it,* by the person standing on the scale. That force is equal and opposite to the force exerted by the scale on the person (action-reaction). This upward force is the normal, **N**. In other words, the scale reading is equal to the magnitude of the normal. If, for some reason, the normal is not equal to the weight (as is the case here), the scale indicates the normal, not the weight. (See chapter five, problem six, for more elaboration on this.)

In the case of the woman standing at the equator it must be true that

$$W - N = F_{NET} = F_c \quad \text{(magnitude only, no vector symbol)}$$

and that $F_c = 4m\pi^2 r/T^2$. Since we know m, r and T, the true weight, W, can be obtained after the value of N is ascertained from the scale reading.

LINK UNKNOWN TO GIVEN

$F_c = 4m\pi^2 r/T^2$ (applied to equator)
$F_c = 4m\ (3.14)^2\ (6.4 \times 10^6)/(8.64 \times 10^4)^2$ (a day in seconds)
$F_c = (3.38 \times 10^{-2})(m)$
$W - 200 = (.038)(W/g)$ $\quad (W = mg)$
$.996W = 200$
$W = 200.8\ N$

The true weight is slightly greater (by less than one percent) than the scale reading, due to the earth's rotation.

(5) Two astronauts in a spacecraft are coasting in a stable circular orbit around, and near, the earth. They plan to "turn on" the engine and burn fuel in order to boost the spacecraft's speed and thereby escape from the earth. If the engine exerts a force of 150,000 N on the 20,000 kg spacecraft, how long must the engine be "on" in order that they achieve their goal?

Gravity

SKETCH AND IDENTIFY

While in orbit a spacecraft continues to move at constant speed *on its own* - without the help of a push provided by an operating engine. The spacecraft moves as a result of its own inertia, and changes direction (in the form of a circle) under the influence of the centripetal force supplied by the earth's gravitational pull. The speed of the spacecraft must be such that the relationship $V_o = \sqrt{gr}$, the formula for orbital speed near earth, is obeyed.

To escape from earth the spacecraft must achieve the minimum speed provided by the relationship $V_e = \sqrt{2GM/R}$. We will assume that the radius of the orbit does not change significantly during the period of time that the engine is operating. In that case, the symbol R in the escape velocity formula can be made equal to the radius of the earth, since the astronauts are orbiting near the earth. (If we don't make this assumption, the problem cannot be solved due to insufficient information regarding R.)

By comparing V_o to V_e we can determine how much speed the spacecraft must gain (Δv). The formula $F = ma$ can be employed to find a, since we know F and m. Then we will be in a position to use the relatioship $a = \Delta v/\Delta t$ to find Δt - the amount of time the engine must act to accelerate the spacecraft in order that the spacecraft attains escape velocity.

LINK UNKNOWN TO GIVEN

$$V_o = \sqrt{gr} = [(9.8)(6.4 \times 10^6)]^{1/2} = 7.92 \times 10^3 \text{ m/s}$$

$$V_e = \sqrt{2GM/R}$$

$$V_e = \left[\frac{(2)(6.67 \times 10^{-11})(6 \times 10^{24})}{6.4 \times 10^6}\right]^{1/2} = 11.18 \times 10^3 \text{ m/s}$$

$\Delta V = V_e - V_o$
$\Delta V = (11.18 \times 10^3) - (7.92 \times 10^3) = 3.26 \times 10^3$ m/s
$F = ma$
$a = 150{,}000/20{,}000 = 7.5$ m/s^2
$a = \Delta v/\Delta t$
$7.5 = 3260/\Delta t$
$\Delta t = 434.67$ sec (7 minutes and 14 seconds)

8.4 ON YOUR OWN PROBLEMS

(Asterisk indicates solution appears in appendices.)

*(1) Due to a major catastrophic explosion deep in the earth's interior, the earth expands to three times its former size. Assuming no change in the earth's mass and ignoring possible changes in the earth's rotational speed, what do you weigh after the explosion if your weight before the explosion was 200 N?

*(2) The planet Uranus has a radius about 3.7 times that of Earth and a mass that is 14.6 times the mass of the earth. If you weigh 100 N on Earth, how much do you weigh on Uranus?

*(3) A particularly important type of orbit around Earth, one used extensively by communications satellites, is known as *geosynchronous orbit*. In that orbit a satellite completes a revolution every 24 hours - the same time it takes the earth to complete a rotation around its axis. How far from the earth's surface must such a satellite be orbiting if the orbit is to be circular and stable?

*(4) What should be the period of the earth's rotation around its axis in order that a person standing on a scale at the equator reads *zero* Newtons?

*(5) What should be the earth's mass in order that objects

moving as fast as light (3 x 10^8 m/s) cannot escape from its surface (assuming no engine to boost the object's speed after it has left the surface of the earth)?

(6) Two platinum spheres are situated near each other with a distance of 40 cm between their centers. The mass of one of them is 1000 kg; the mass of the other is 400 kg. (a) How strong is the gravitational attraction of the smaller one toward the larger one? (b) How strong is the attraction of the larger one toward the smaller one? (c) What would be the initial acceleration rate of the smaller one toward the larger one if frictional effects were completely eliminated?

(7) If the earth were suddenly to shrink to one-fifth its current size, what would be your weight if you now weigh 200 N?

(8) The planet Jupiter's radius is about 11.2 times that of Earth and its mass is 318 times the mass of the earth. If you weigh 100 N on Earth, how much do you weigh on Jupiter?

(9) What is the period of a spaceship in a stable circular orbit around and near the earth?

(10) What is the radius of a stable circular orbit around the earth if the speed of rotation is 2 km/sec?

(11) At what point between the earth and the moon are the gravitational forces they exert equal in magnitude? (The moon's mass is one-eightieth that of the earth and its diameter is one-fourth the diameter of the earth.)

(12) What is the period of rotation of a spacecraft in a stable circular orbit around the earth at a distance of two earth-radii from the earth's center?

(13) What is the acceleration rate of any freely falling object on Jupiter? (Use the data provided in problem eight.)

(14) What is the velocity of escape from the surface of the earth in km/hr?

(15) What is the velocity of escape from the surface of the moon? (Use the data provided in problem eleven.)

(16) Is the following statement correct? "The apple falls downward toward the earth instead of the earth rising upward to meet the apple, because the earth has much more mass and consequently exerts the greater pull." Explain your answer. If the statement is incorrect, explain why the apple is indeed seen to be moving toward the earth while the earth does not appear to be moving toward the apple.

(17) In figure 8-5 the ball at point C contains 4 kg of mass and the spheres at A and B contain 20 kg each. Determine the direction and magnitude of the resultant gravitational force on, and the acceleration rate of, the ball at point C.

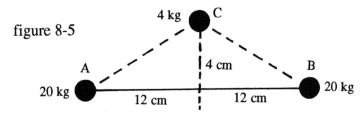

figure 8-5

(18) You land on a planet where all freely falling objects accelerate at the rate of 4 m/s^2. You criss-cross the planet's surface and determine its circumference to be 20,000 km. What is the mass of the planet?

(19) Give two reasons why your scale reading (the reading obtained when standing on a scale) is lower at the equator than at the north pole. (Hint: one reason is based on the shape of the earth.)

(20) Two satellites, A and B, revolve around the earth in stable circular orbits at the same distance from the earth's surface. The

mass of satellite A is nine times the mass of B. If satellite A is moving at the rate of 2000 m/s, how fast is satellite B moving?

(21) One of Jupiter's many moons revolves around Jupiter at four times the distance of another moon. If the period of the moon with the smaller orbital radius is four months, what is the period of the moon with the larger orbital radius?

9

CHAPTER NINE

ELECTRICITY

9.1 VOCABULARY AND FORMULAS

WORD	SYMBOL	UNIT
Charge	Q, q	Coulomb, C
Intensity	E	N/C

DEFINITIONS

CHARGE (positive and negative) - Names of entities that exist in the universe other than matter.

ELECTRIC FORCE - Like charges repel; unlike charges attract.

ELECTRIC FIELD - Entity that permeates the space in the vicinity of a charge. The field acts to exert a force on any other charge situated in the field.

ELECTRIC FIELD INTENSITY, **E**, direction of - Defined as the direction of the force the field would exert, at any particular point in the field, on a positive charge situated at that point.

ELECTRIC FIELD INTENSITY, **E**, magnitude of - Equal to the magnitude of the force the field would exert, at any particular

Electricity

point in the field, on a charge of one coulomb situated at that point. Since the magnitude of the force, F, exerted by a field on a charge at a particular point in the field is proportional to the amount of charge, q, located at that point (Coulomb's law), the formula $F = Eq$ must be valid. The formula is also valid in vector form, $\mathbf{F} = \mathbf{E}q$, provided we are careful with the sign of q - making it positive when the charge acted upon by the field is positive, and negative when that charge is negative.

FIELD LINES - Lines or curves with arrows superimposed upon them, sketched or imagined to exist in the vicinity of a charge or group of charges, that reveal the pattern of the electric field intensity (direction and magnitude) in various regions of the field. The arrows on the field lines indicate the direction of the electric field intensity, and the concentration of the lines indicates the magnitude of the field intensity. As the direction and magnitude of the field intensity vary from point to point in the field, so do the orientation and concentration of the lines. Where the lines are more concentrated, the field is proportionally more intense.

COULOMB - The unit of charge. Defined as the amount of charge which when separated from an equal amount of charge by one meter exerts a force of 9×10^9 Newtons on the other charge (assuming both are *point* charges).

TEST CHARGE - Charge placed at a point to test for the presence of an electric field at that point. If an electric field is present at the location of the charge, the charge experiences an electric force; if no such field exists at that point, no (electric) force appears.

FORMULAS

a. $F = KQ_1Q_2/d^2$ $\quad K = 9 \times 10^9$ N·m²/C² $\quad$ **b.** $\mathbf{F} = \mathbf{E}q$, $\quad F = Eq$

9.2 THINGS TO KNOW

(1) Coulomb's law, $F = KQ_1Q_2/d^2$, provides the magnitude of the electric force exerted by either of two point charges on the other. The electric forces exerted by each charge on the other are equal in magnitude and opposite in direction and act parallel to the straight line connecting the charges (figure 9-1).

figure 9-1   {$F_1 = -F_2$ in all situations}

(2) Coulomb's law is designed for point charges - situations where the charge on each of two objects is distributed over an area that is small compared to the distance between the objects. When this is not the case, Coulomb's law cannot be used. The only exception is the case of large spherically distributed charges of uniform density, where Coulomb's law may be used provided d represents the distance between the centers of the spheres. Coulomb's law cannot be used, for example, in the important case of oppositely charged parallel plates (known as a *capacitor*) because the dimensions of the plates are not small compared to the distance between them.

(3) Field lines are *lines of force*. They do not indicate the path taken by a charge (positive or negative) in the field. That path depends, among other things, on the motion of the charge (direction and speed) as it enters the field. The field lines do indicate the direction of the *net electric force* exerted on a positive charge at various points in the field. (Recall that force and motion are not

Electricity 197

synonomous - chapter five, Things to Know, item one.)

(4) Field intensity, **E**, is a vector quantity. Where two or more fields overlap, the net field intensity at any point is the vector sum (resultant) of the individual field intensities (each created by a different charge) at that point.

(5) Be careful when using terms such as "charge density", "concentration of charge", "field line density" and "concentration of field lines". There are two types of density (or concentration) - density per volume and density per surface area. It is the density of field lines *per unit surface area* (with the surface perpendicular to the lines) that indicates the magnitude of the intensity of the electric field, not the density per volume. The former is expressed as a number of lines per square meter, the latter as a number of lines per cubic meter.

When a particular amount of charge is placed on a conductor, the charge distributes itself on the surface of the conductor. The density of charge per *square meter of surface area* is uniform throughout the surface of the conductor, but the density of charge per *cubic meter of volume* is not necessarily uniform - it is greatest in the vicinity of corners and sharp points, if any exist.

(6) The symbol q in the formula $\mathbf{F} = \mathbf{E}q$ refers to the charge upon which the electric field acts (known as the *test charge*), not the charge setting up the field (which is usually designated by the symbol Q).

(7) As a consequence of what has been said above regarding field lines, the following statements turn out to be true:

A. Field lines never intersect each other.

B. Near a positive charge, the arrows point away from the charge; near a negative charge, the arrows point toward the charge.

C. Field lines begin and end on charges, not in empty space.

(8) The following are some often seen and used field line representations:

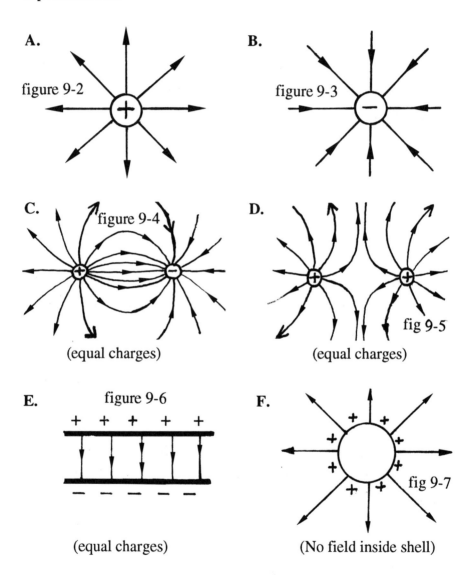

figure 9-2
figure 9-3
figure 9-4 (equal charges)
fig 9-5 (equal charges)
figure 9-6 (equal charges)
fig 9-7 (No field inside shell)

Note: In all these field representations it is understood that the field lines continue either to infinity or until the loops become closed, except where the lines encounter charge. The field representations provide a pattern that are to be interpreted imaginatively.

Electricity

9.3 SOLVING THE PROBLEMS

(1) What is the magnitude of the electric field intensity at a point 4 meters distant from a 3 C charge?

SKETCH AND IDENTIFY

figure 9-8

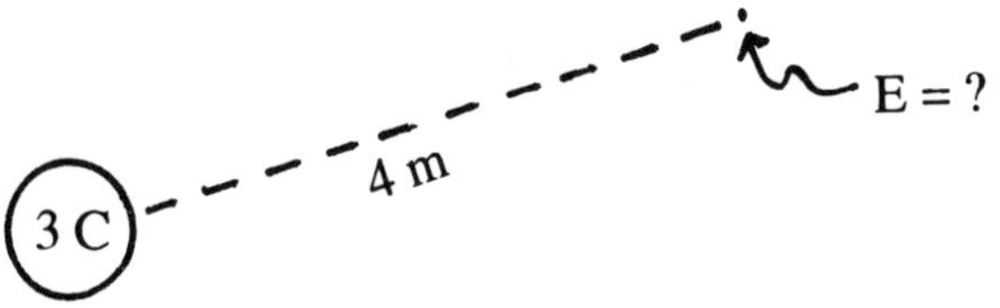

3 C ---- 4 m ---- E = ?

The definition of the magnitude of **E** as the magnitude of the electric force exerted on one coulomb, can be used here to determine the answer to our question. We simply make believe that one coulomb of charge has been placed at the point of interest and apply Coulomb's law to determine the force exerted on that imaginary one coulomb by the three coulomb charge. The magnitude of that force is the magnitude of the intensity of the field, E, at that point (method one below).

Alternatively, we can make believe that any charge, q, has been placed at the point of interest. To distinguish this charge from the other one, the three coulomb charge is labelled Q. From the equation $F = Eq$ we know that $E = F/q$, where q is the imaginary test charge at the point of interest. Knowledge of the ratio F/q then leads to knowledge of the magnitude of **E** (method two below).

LINK UNKNOWN TO GIVEN
method one

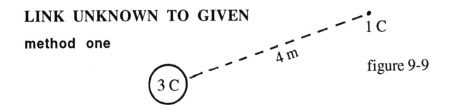

figure 9-9

$F = KQ_1Q_2/d^2 = (9 \times 10^9)(3)(1)/4^2 = 1.69 \times 10^9$ N

Therefore, by definition, E = 1.69×10^9 N/C

method two

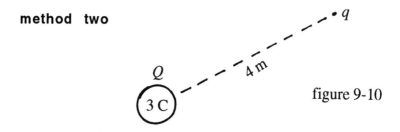

figure 9-10

$E = F/q = (KQq/d^2)/q = KQ/d^2 = (9 \times 10^9)(3)/4^2 = 1.69 \times 10^9$ N/C

(2) Two charges, one positive, the other negative, are separated by six centimeters. Each of them carries 6μC of charge (six micro-coulombs, the prefix *micro* meaning one-millionth or 10^{-6}). What is the direction and magnitude of the net electric field intensity at a point 6 cm distant from both charges?

SKETCH AND IDENTIFY

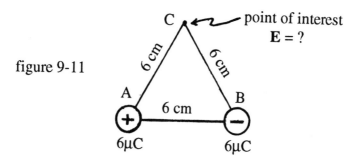

figure 9-11

The net **E** vector at the point of interest (point C in figure 9-11) is the resultant of the **E** vector set up there by the charge at point A and the **E** vector set up there by the charge at point B. These **E** vectors can be determined by imagining that one coulomb of positive charge has been placed at the point of interest, then applying Coulomb's law with its associated rules to find the force vector exerted by each 6μC charge on the imaginary one coulomb charge.

To find the force vector exerted by the positive 6μC charge (at point A in figure 9-11) on the imaginary positive charge of one coulomb (at point C) use will have to be made of the diagram in figure 9-12. Since the two 6μC charges (at points A and B) and the point of interest (point C) form an equilateral triangle, each of the angles of the triangle must be 60°. Since the angle between vector F_1 (the force exerted by the positive 6μC charge on the imaginary one coulomb charge) and the horizontal must be equal to the angle between the horizontal and line AC (they are corresponding angles formed by two parallel lines intersected by a transversal), we conclude that F_1 is oriented 60° from the horizontal (figure 9-12).

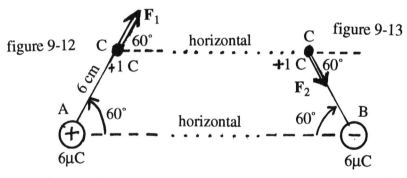

To find the force vector exerted by the negative 6μC charge (at point B in figure 9-11) on the imaginary positive charge of one coulomb, use will have to be made of the diagram in figure 9-13. The angle between vector F_2 (the force exerted by the negative 6μC charge on the imaginary charge of one coulomb) and the horizontal also is 60° since it must be equal to the angle between the horizontal and line BC (they are alternate interior angles formed by two parallel lines intersected by a transversal).

LINK UNKNOWN TO GIVEN

$F_1 = KQq/d^2 = (9 \times 10^9)(6 \times 10^{-6})(1)/(6 \times 10^{-2})^2$ (6 cm = .06 m)

$F_1 = 1.5 \times 10^7$ N

$F_2 = KQq/d^2 = (9 \times 10^9)(6 \times 10^{-6})(1)/(6 \times 10^{-2})^2 = 1.5 \times 10^7$ N

Since both force vectors are exerted on one coulomb of positive charge, they are essentially **E** vectors. Thus E_1 and E_2 at the point of interest are arranged as in figure 9-14.

The resultant of these two vectors, E_{NET}, can easily be ascertained from their components.

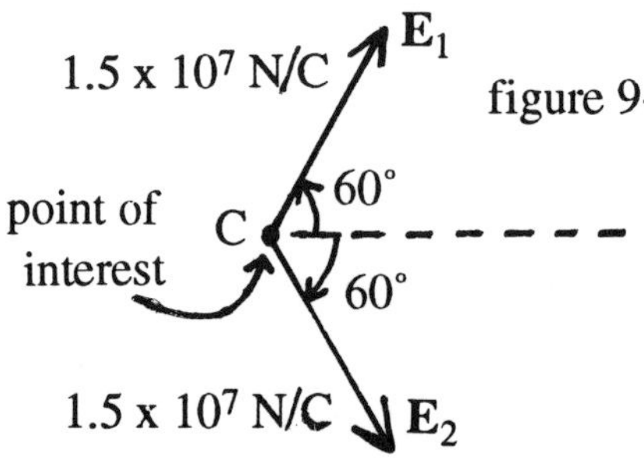

$E_{NET} = E_1 + E_2$

$E_{NET,Y} = E_{1Y} + E_{2Y}$

$E_{NET,X} = E_{1X} + E_{2X}$

figure 9-14

Since the two y components are equal in magnitude (each is 1.5×10^7 times the sine of 60°) and opposite in direction, their vector sum is zero. The y component of E_{NET} is, therefore, zero. The two x components, on the other hand, point in the same direction (to the right), each being equal to 1.5×10^7 times the cosine of 60°. Thus:

$E_{NET,X} = (2)(1.5 \times 10^7) \cos 60° = 1.5 \times 10^7$ N/C.

The resultant electric field vector, E_{NET}, at the point of interest is therefore directed horizontally to the right, with a magnitude of 1.5×10^7 N/C.

> Note: This result agrees with our earlier depiction of the field in the vicinity of two charges, one positive, the other negative (Things to Know, item 8c, figure 9-4). At all points on an imaginary vertical line drawn midway between the charges, the field lines are oriented horizontally, directed away from the positive charge, toward the negative charge.

Electricity

(3) A negatively charged oil droplet remains suspended between two oppositely charged, parallel plates (figure 9-15). The droplet's mass is .01 gram and it carries a charge of 2×10^{-5} C. (a) What is the magnitude of the electric field intensity at the point where the droplet hangs? (b) If the droplet is moved two centimeters closer to the lower, negatively charged plate, would it still remain suspended (assuming it is at rest when placed at the new position)?

```
          +   +   +   +   +   +
        ─────────────────────────
                    ⊖
figure 9-15

        ─────────────────────────
          ─   ─   ─   ─   ─   ─
```

SKETCH AND IDENTIFY

To "remain suspended" is to be in a state of equilibrium. This implies that the electric force directed upward is balanced by the droplet's weight directed downward (figure 9-16). From the droplet's given mass we can calculate its weight (via $W=mg$) and from its weight the electric force can be determined. Since the charge on the droplet is known, we can then make use of the relationship $F=Eq$ to find the magnitude of the electric field intensity.

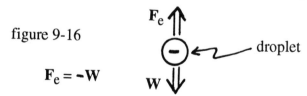

figure 9-16

$F_e = -W$

The pattern of field lines between two such plates (Things to Know, item 8e, figure 9-6) enables us to compare the field intensity at various points. We should be able to ascertain from that field pattern whether the intensity at the new location (two cm closer to the negative plate) is greater than, equal to, or smaller than its value at the original location. This, in turn, should reveal whether the droplet remains suspended, falls down, or accelerates upward, when placed at the new location.

LINK UNKNOWN TO GIVEN

(a) $W = F_e$ (magnitude only)

$W = mg$ and $F = Eq$

$mg = Eq$

$(1 \times 10^{-5})(9.8) = (E)(2 \times 10^{-5})$ $(.01 \text{ gm} = 1 \times 10^{-5} \text{ kg})$

$E = 4.9$ N/C

(b) Since the field lines between the plates run parallel to each other, their concentration there is uniform (except near the edges). Consequently, the electric field intensity must also be uniform between the plates. (The reason for this uniformity is partly based on the fact that points closer to one plate are farther from the other plate.) Nothing, therefore, changes at the new location - not E, m, g or q. The droplet therefore remains suspended at the new location, for the same reason that it remained suspended at the original location. The gravitational and electric forces are the same as before - equal and opposite to each other.

(4) A one gram ping-pong ball hangs at the end of a non-conducting string. This ball and another one are then charged equally. When the second ball is brought near the suspended ball, the latter is deflected by an electric force of repulsion until the string makes an angle of 30° with the vertical. The string thereafter remains so oriented as long as the second ball is held at a distance of 4 cm from the first (figure 9-17). What amount of charge does each ball carry?

Electricity 205

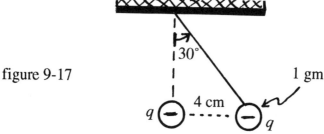

figure 9-17

SKETCH AND IDENTIFY

As the string hangs there at an angle of 30° from the vertical, a state of equilibrium must exist between the forces of gravity, tension and electricity that act on the suspended ball. These forces are oriented as in figure 9-18. The vector equation **T**+**F**$_e$+**W** = **0** must therefore be valid in that position.

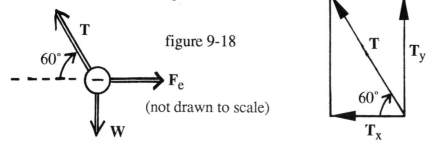

figure 9-18

(not drawn to scale)

We can solve the above equation for **F**$_e$ in a series of steps, starting with knowledge of **W** (the mass of the ball is given and **W**=mg). For equilibrium to exist, it must exist in the vertical dimension and independently in the horizontal dimension. Knowledge of **W** therefore leads to knowledge of **T**$_y$ (the vertical, upward component of **T**) - they must be equal and opposite in order that equilibrium exist in the vertical dimension - and knowledge of **T**$_y$ leads, in turn, to knowledge of **T**$_x$ - their magnitudes are linked trigonometrically as components of the same vector. From the magnitude of **T**$_x$ can be obtained the magnitude of **F**$_e$, since these forces must be equal and opposite in order that equilibrium exist in the horizontal dimension. Knowledge of the magnitude of F_e can then be combined with Coulomb's law to determine the charge, q, on each ball.

LINK UNKNOWN TO GIVEN

$T + F_e + W = 0$

$T_x + F_{ex} + W_x = 0$ (equilibrium in the horizontal dimension)

$T_y + F_{ey} + W_y = 0$ (equilibrium in the vertical dimension)

Since W is directed vertically downward, $W_x = 0$ and $W_y = mg$ (magnitude only). Since F_e is directed horizontally to the right, F_{ey} is equal to *zero* and $F_{ex} = F_e$ (magnitude only). We can therefore write:

$T_x = F_e$ and $T_y = mg$ (magnitude only)

$mg = (1 \times 10^{-3})(9.8) = 9.8 \times 10^{-3}$ N (one gram = .001 kg)

$T_y = 9.8 \times 10^{-3}$ N

$T_y/T_x = \tan 60°$ (see figure 9-18)

$1.73 = 9.8 \times 10^{-3}/T_x$

$T_x = 5.66 \times 10^{-3}$ N

$F_e = T_x = 5.66 \times 10^{-3}$ N

Now, the electric force, F_e, exists as a result of the presence of the two equal charges that are separated by 4 cm. Therefore:

$F_e = KQ_1Q_2/d^2$ $Q_1 = Q_2$

$5.66 \times 10^{-3} = (9 \times 10^9)Q^2/(4 \times 10^{-2})^2$ (4 cm = .04 m)

$Q^2 = (16 \times 10^{-4})(5.66 \times 10^{-3})/9 \times 10^9 = 10.06 \times 10^{-16}$

$Q = 3.17 \times 10^{-8}$ C

Electricity 207

9.4 ON YOUR OWN PROBLEMS

(Asterisk indicates solution appears in appendices.)

*(1) If all the electrons in one gram of ordinary water are placed on the north pole of the earth and all the protons are placed at the equator, how strong is the electric force exerted by either group of charges on the other? (Every electron and proton carries a charge of 1.6×10^{-19} C, one gram of H_2O has 1/18 of a mole molecules, one mole is 6×10^{23}, every oxygen atom contains 8 protons and 8 electrons and every hydrogen atom has 1 proton and 1 electron.)

*(2) Two one-gram pith balls hang side by side at the end of one meter long, non-conducting threads. When they are both charged equally, the pith balls repel each other until the threads remain 6° apart. How much charge does each pith ball carry?

*(3) An *electric dipole* consists of a pair of equal charges, one positive, the other negative. Show that at large distances from the dipole (measured perpendicularly to the line connecting the charges) the net electric force exerted by the dipole varies, as a good approximation, with the inverse of the cube (third power) of the distance to the dipole.

*(4) An electron (9.1×10^{-31} kg, 1.6×10^{-19} C) is projected into the region between two oppositely charged plates (figure 9-19). The electron enters horizontally at a point midway between the plates with a speed of 3×10^6 m/s. The magnitude of the electric field intensity between the plates is 10^3 N/C, the plates are 8 cm apart and 6 cm long. (Assume the plates are large enough and close enough for the field to be uniform, as illustrated in figure 9-6). A vertical screen is situated 20 cm from the plates, as in figure 9-19. (a) By how many degrees will the final path of the electron, as it approaches the screen, be deflected from its original, horizontal

direction? (b) How many centimeters below its original destination does the electron strike the screen?

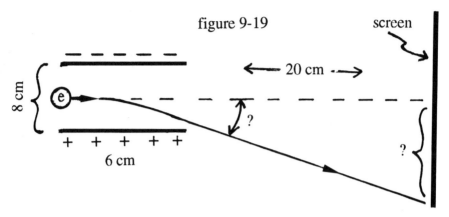

figure 9-19

(5) How strong is the electric force of attraction between the single proton in the nucleus of a hydrogen atom and the electron that orbits around it? The radius of a hydrogen atom is about 5×10^{-11} meters.

(6) Each of two equal positive charges is situated on a different corner of an equilateral triangle whose sides are 4 cm long. If each charge carries 2×10^{-4} C, what is the direction and magnitude of the net force they exert on a positive charge of 3×10^{-3} C situated on the third corner of the triangle?

(7) Assume that the electric field intensity between the charged plates in problem four can be varied and that an alpha particle is projected into the region between the plates. How intense must the field be in order that the angle of deflection is 20°? (An alpha particle consists of two protons and two neutrons; each proton and neutron has a mass of 1.67×10^{-27} kg.)

(8) If the intensity of an electric field at a point outside and two meters distant from a charged sphere (measured from the center) is 1000 N/C, how much charge does the sphere carry?

Electricity

(9) An electron is situated inside a charged spherical shell of radius 8 cm, at a point one cm distant from the shell. The shell carries 4×10^{-6} C of charge. What is the magnitude and direction of the electric force exerted on the electron?

(10) If the electron in problem nine is now moved to a point one cm distant from the shell but outside the shell, what is the magnitude and direction of the electric force exerted on the electron?

(11) How much charge must be placed on a .5 gm pith ball in order that it remain suspended between two oppositely charged parallel plates with a field intensity of 20 N/C?

(12) A metal sphere carries a net charge of $-4\mu C$. How many electrons does the sphere have in excess of protons?

(13) The two protons inside a helium nucleus are separated (center to center) by a distance of about 10^{-15} meter. How strong is the force of repulsion between them?

(14) The electric field intensity at a particular location in the vicinity of a point charge is 20 N/C. How intense is the field at a point twice as far from the charge?

(15) An electron is released from rest at the inside surface of the negatively charged plate of a capacitor (figure 9-6). One thousandth of a second later it strikes the positively charged plate three cm away. (a) What is the electric field intensity between the plates? (b) How fast is the electron moving as it strikes the positive plate? (The mass of an electron is 9.1×10^{-31} kg.)

(16) Near the surface of the earth there exists an electric field whose intensity is 100 N/C directed downward. How many extra electrons must be placed on a dime of mass 4 gms in order that the electric force just balances the weight of the dime and the dime remains suspended in mid-air?

10

CHAPTER TEN

MAGNETISM

10.1 VOCABULARY AND FORMULAS

WORD	SYMBOL	UNIT
Current	I	Coulombs/sec (C/s),
Current	I	Ampere (A)
Intensity	**B**	Newton/Amp·meter
Induction	**B**	Tesla (T)
Flux density	**B**	Webers/meter^2 (W/m^2)

DEFINITIONS

CURRENT - Rate of flow of charge through a conducting material, usually expressed in amperes. One ampere is equal to the passing of one coulomb of charge past any particular point in the conductor, per second.

MAGNETIC FIELD - Entity that permeates the space in the vicinity of a moving charge. This field exerts a force on any other charge that moves in the field.

MAGNETIC FIELD INTENSITY, **B**, direction of - Determined by the moving charge that sets up the field via hand rules one and

Magnetism 211

two (described below) and various formulas. Same as direction of arrow drawn from the *S*-pole to the *N*-pole of a test magnet placed at any particular point in the field, after the magnet settles into its equilibrium orientation.

MAGNETIC FIELD INTENSITY, **B**, magnitude of - Equal to the magnitude of the force the field would exert on a one meter long wire carrying a current of one ampere. Also equal to the magnitude of the force the field would exert on one coulomb of charge moving, at any particular point in the field, at the rate of one meter per second. The magnitude of a magnetic field's intensity at any point in the field is decided, as is the direction, by the moving charge that sets up the field.

MAGNETIC FIELD INDUCTION - Same as magnetic field intensity.

MAGNETIC FORCE - Force between moving charges, independent of the electric force (which exists even when the charges are not moving). The rule, "like poles repel, unlike poles attract", is *not* a general description of the magnetic force, merely one of various manifestations of that force.

MAGNETIC FLUX - Fancy name for magnetic field lines, which are also referred to as *flux lines*. Each flux line is also called a *Weber*. The arrows on magnetic field lines indicate the direction of the intensity of the field at any particular point in the field.

FLUX DENSITY - Number of flux lines per square meter of surface oriented perpendicularly to the lines (thus the term *Webers per square meter*). This density is imagined to be equal to the magnitude of the field's intensity at any particular point in the field. Where the field is more intense, the density of field lines is greater, and the lines are more crowded together.

$\boxed{\text{X X X X}}$ - Symbol for "perpendicular to page, directed *away from* the reader". In other words, *into the page*.

$\boxed{\cdot\ \cdot\ \cdot\ \cdot}$ - Symbol for "perpendicular to page, directed *toward* the reader". In other words, *out of the page*.

FORMULAS

a. $F = BIL$ **b.** $F = Bqv$ **c.** $F = \mu I_1 I_2 L/d$ $\mu = 2 \times 10^{-7}$ N/A^2
d. $B = \mu I/d$

e. Hand rule #1, to find direction of magnetic field intensity in the vicinity of a straight line current of charge: If the moving charges are negative, grasp (or imagine that you're grasping) the current-carrying wire with your left hand, with the outstretched thumb pointing in the direction of electron flow. The four fingers then indicate the direction of the arrows to be placed on the circular field lines that are wrapped around and centered on the current, with the plane of each circle perpendicular to the current (figure 10-1). If the current consists of moving positive charges, the right hand is used.

EXAMPLE

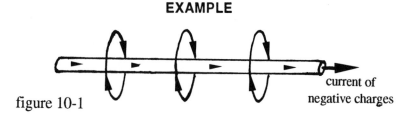

figure 10-1

current of negative charges

In front of the wire, the field is directed toward the top of the page.
Behind the wire, the field is directed toward the bottom of the page.
Above the wire, the field is perpendicular to the page, away from you.
Below the wire, the field is perpendicular to the page, toward you.

f. Hand rule #2, to find direction of magnetic field intensity in the vicinity of a circular current of charge: If the moving charges are

Magnetism

negative, bend and orient the four fingers of your left hand so that the fingertips point in the direction of the current (either clockwise or counter-clockwise). The outstretched thumb then indicates the direction of the field intensity inside the circular current, including the region in front of and behind the area enclosed by the circle (figure 10-2). The field intensity's direction is the reverse of this outside the circle of current. If the current consists of moving positive charges, the right hand is used.

EXAMPLE

figure 10-2

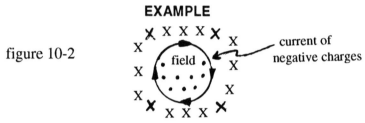

g. Hand rule #3, to find direction of force (not field!) exerted by a magnetic field on a charge or group of charges moving in the field: The motion of the charge, or at least a component of the motion, must be perpendicular to the direction of the field's intensity at the point where the charge is moving, otherwise rule #3 does not apply and no force is exerted. If the moving charge is negative, point the index finger of your left hand in the direction of the field intensity (at the point where the charge is moving). Then, with the thumb perpendicular to the index finger, point the thumb in the direction of motion of the charge (or in the direction of the component of the motion that is perpendicular to the field). Finally, orient the middle finger of your left hand perpendicularly to the palm. The middle finger now indicates the direction of the force exerted by the field on the moving negative charge (figure 10-3). If the moving charge is positive, the right hand is used. The hand, wrist, palm and index finger must form a straight line when using this rule.

EXAMPLE

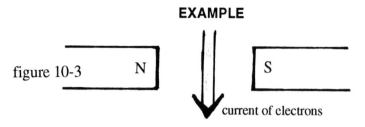

figure 10-3

1. Index finger to the right, in the direction of field intensity. ⟶
   (Same as direction of arrow drawn from S to N pole (S⟶N) of a test magnet placed in the field, at equilibrium.)
2. Thumb downward, in the direction of motion of negative charge. ↓
3. Middle finger into the page, indicates direction of force exerted by field on current. ×××

10.2 THINGS TO KNOW

(1) The essence of magnetism and the magnetic force is not "magnets act on other magnets". Instead, the magnetic force is best summarized by the following statement: *Moving charges act on other moving charges.* Various formulas and hand rules help fill in the details of this force - how strong it is and in what direction it acts. The action of permanent magnets on each other (like poles repel, unlike poles attract) is incorporated into magnetism by virtue of the fact that the behavior of magnets is based on their internal structure as coil shaped currents (*solenoids*).

(2) Magnetic field lines are not "force lines" (unlike electric field lines, which *are* "lines of force"). The direction of the magnetic force is never the same as the direction of the magnetic field lines. Force lines cannot be drawn for magnetism (as they are for electricity) because the direction of the magnetic force depends on the motion of the charge it acts upon - something that is not under the exclusive control of the field. In other words, the direction of the

Magnetism 215

field lines do not determine the direction of the force exerted by the field.

(3) Care must be taken in using hand rule three not to confuse the motion of a charge that *leads to* the exertion of a magnetic force with the motion that is a *consequence* of that force. The thumb in hand rule three must point in the direction of the motion that leads to the exertion of a magnetic force (whose direction is indicated by the middle finger). Don't be lulled into the habit of loosely associating the thumb with "current". To do so is to invite much confusion. Sometimes the current is a consequence of (instead of the cause of) the exertion of a magnetic force (as is the case with current induced by a generator) and is to be associated with the middle finger - which represents that force - and not the thumb. In such cases, the movement of a conducting wire across the field lines is the motion that leads to the exertion of the magnetic force, and it is this movement that is associated with the thumb. In other situations, such as the case of a battery driven current sitting in a magnetic field, the current is the motion that leads to the exertion of a magnetic force (on the wire) and is properly associated with the thumb. Simply put, don't associate the thumb with "current"; instead, associate the thumb with the *motion of the charge that causes the field to exert a force*. That motion may consist of a flow of electrons within a wire or the movement of a wire across magnetic field lines.

(4) Before using either $F = Bqv$ or $F = BIL$ verify that the direction of **v** or **I** is perpendicular to **B**, or that at least a component of **v** or **I** is perpendicular to **B**. The formulas cannot be used, and if used lead to incorrect results, when **v** or **I** are parallel to **B** (no force exists in that case). When only a component of **v** or **I** is perpendicular to **B**, it is the magnitude of that component that gets plugged

into the symbol v or I in these formulas.

(5) Analogously, before using the formula $F = \mu I_1 I_2 L/d$ verify that the currents are parallel to each other, or that at least a component of one of them is parallel to the other. The formula is not applicable to currents that are perpendicular to each other. There is no force in such situations. If only a component of I_2 is parallel to I_1, it is the magnitude of that component that takes the place of I_2 in the equation. The symbol L in the formula represents the length of the parallel arrangement between the currents, and d represents the distance between the currents.

(6) The unit of B is expressed in a variety of forms in the literature, but all represent the same thing - the magnitude of the intensity of the magnetic field. The N/A·m, the Tesla and the W/m² all represent one unit of magnetic field intensity (B) in the MKS system. The magnitude of **B** is also associated with a variety of labels, such as *magnetic induction* and *flux density*. Don't permit these fancy phrases to intimidate you. They all merely represent different ways of looking at the same thing.

(7) Two parallel currents in the same direction, attract each other; in opposite directions, they repel each other. This pattern is the reverse of the relationship that exists between charges and poles, where "likes" repel and "unlikes" attract.

(8) The formula $B = \mu I/d$ provides the magnitude of the intensity of the magnetic field created by a long, straight line current at a point d meters distant from the current (figure 10-4).

figure 10-4

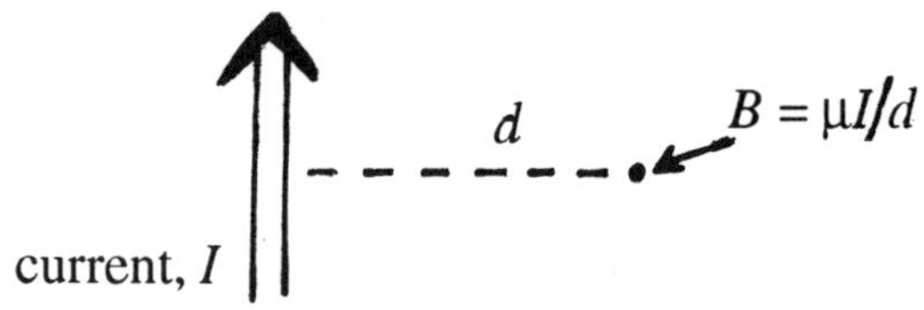

current, I

$B = \mu I/d$

Magnetism

(9) The force between currents, whose magnitude is provided by the formula $F = \mu I_1 I_2 L/d$, is one of many manifestations of the magnetic force. The first current (either current can be labeled "first" and the other "second") sets up a magnetic field (as any moving charge does) and that field acts on the second current (as any magnetic field acts on moving charges). The magnitude of the intensity of the magnetic field set up by the first current is provided by the expression $B = \mu I_1/d$ (Formulas, item d). The force exerted by this magnetic field on the second current is provided by the equation $F = BI_2L$ (item a). Combining these two formulas leads to $F = \mu I_1 I_2 L/d$ - the formula for the force between two currents.

10.3 SOLVING THE PROBLEMS

(1) Two currents of electrons, one of 2 amperes, the other of 4 amperes, are arranged parallel to each other as in figure 10-5. The wires carrying the currents are separated by 8 cm and the parallel arrangement is 3 meters long.

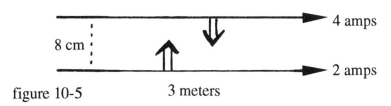

figure 10-5

(a) Show that the field created by the 2-amp current acts to pull the wire carrying the 4-amp current downward. (b) Show that the field created by the 4-ampere current acts to attract the 2-ampere current upward. (c) How strong is the force exerted by either current on the other? (d) If the 4-amp current is replaced by a single proton traveling in the same direction at the rate of 6 x 10³ m/s, what is the magnitude and direction of the force exerted on the proton?

SKETCH AND IDENTIFY

The direction of the magnetic field intensity created by either current at the location of the other is determined by hand rule one. The direction of the force exerted by that field on the other current is then determined by hand rule three.

The magnitude of the force exerted by either current on the other can be determined from the formula $F = \mu I_1 I_2 L/d$. Alternatively, the magnitude of the magnetic field intensity is first found by applying $B = \mu I/d$ and the force is then determined from $F = BIL$.

To find the force exerted by the 2-amp current on the moving proton we must first know the magnetic field intensity (magnitude

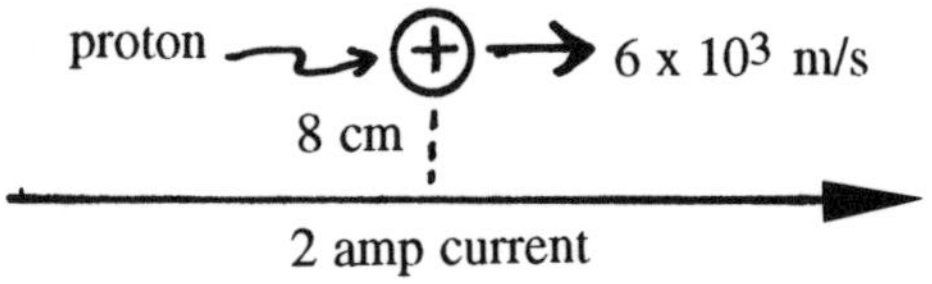

```
                proton  ⤳ ⊕ ⟶ 6 x 10³ m/s
figure 10-6             8 cm
           ─────────────────────────────▶
                     2 amp current
```

and direction) created by the 2-amp current at the location of the proton (figure 10-6). Then hand rule three can be employed to find the direction of the force and the formula $F = Bqv$ is applied to find the magnitude of the force.

LINK UNKNOWN TO GIVEN

(a) The field created by the 2-amp current at the location of the 4-amp current is directed as in figure 10-7. This can be ascertained by resorting to hand rule one (Formulas, item e). The 4-amp current

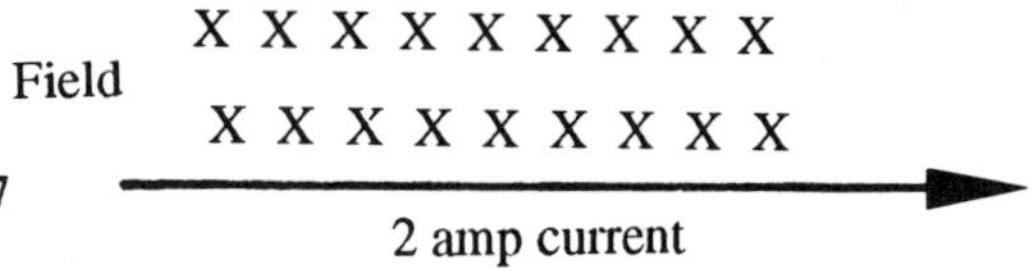

```
                 X X X X X X X X X
         Field
                 X X X X X X X X X
figure 10-7  ─────────────────────────────▶
                     2 amp current
```

Magnetism   219

is therefore situated in a magnetic field that is directed perpendicularly to the page, away from the reader, as in figure 10-8.

```
                 x x x x x x x x x x x
       ─────────────────────────────────────▶
                 x x x x x x x x x x x         4 amperes
   figure 10-8
```

This field acts on the 4-amp current in the manner dictated by hand rule three:

        index finger - field intensity: [x x x]
        thumb - motion of negative charge: →
        middle finger (left hand) - force: ↓

(b) The field created by the 4-amp current at the location of the 2-amp current is oriented as in figure 10-9. The 2-amp current is

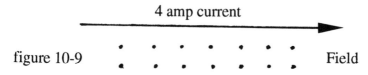

figure 10-9

therefore situated in a magnetic field that is directed perpendicularly to the page, toward the reader, as in figure 10-10.

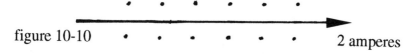

figure 10-10   2 amperes

This field acts on the 2-amp current as follows:

        index finger - field intensity: [· · ·]
        thumb - motion of negative charge: →
        middle finger (left hand) - force: ↑

Since the upper current is pulled down and the lower current is pulled up, it certainly looks like "two currents attract each other when the charges flow in the same direction".

(c) The force either current exerts on the other is provided by:

$F = \mu I_1 I_2 L/d = (2 \times 10^{-7})(4)(2)(3)/(8 \times 10^{-2})$ (8 cm = .08 m)

$F = 6 \times 10^{-5}$ N

(d) When a proton replaces the 4-amp current, we have the following arrangement (figure 10-11):

```
                    X X X   proton   X X X X
figure 10-11        X X X    (+)→    X X X X        Field
                    X X X X X X X X X X X
                    ─────────────────────────────→
                             2 amp current
```

The magnitude of the magnetic field created by the 2-amp current at the location of the proton is determined as follows:

$B = \mu I/d = (2 \times 10^{-7})(2)/(8 \times 10^{-2}) = 5 \times 10^{-6}$ N/A·m

This field exerts a force on the moving proton whose magnitude is:

$F = Bqv = (5 \times 10^{-6})(1.6 \times 10^{-19})(6 \times 10^{3})$

$F = 4.8 \times 10^{-21}$ N

The direction of the force is determined via hand rule three:

        index finger - field intensity: ⊠⊠⊠

        thumb - motion of positive charge: →

        middle finger (right hand) - force: ↑

(2) A proton is projected into the magnetic field shown in figure 10-12. Describe the path taken by the proton if it is projected with a

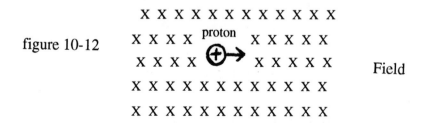

figure 10-12

velocity of 2×10^6 m/s and the magnitude of the field intensity is 0.5 N/A·m. (Proton: charge - 1.6×10^{-19} C, mass - 1.67×10^{-27} kg).

SKETCH AND IDENTIFY

As the proton begins moving to the right in figure 10-12 the force exerted on it by the magnetic field is directed upward, in the plane of the page (hand rule three, using the right hand). As a result,

figure 10-13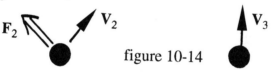

the proton will - after some time - be moving in a northeasterly direction (figure 10-13). But as the proton's direction of motion changes, so does the direction of the magnetic force exerted on it, since that force must always be perpendicular to the motion. We now get (figure 10-14):

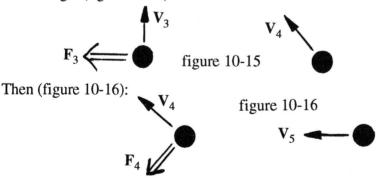

figure 10-14

Then we get (figure 10-15):

figure 10-15

Then (figure 10-16):

figure 10-16

The proton, which used to be moving in the direction of v_1 (to the right), has its direction of motion change to v_2, then to v_3 (upward), then to v_4, then to v_5 (leftward). This process continues as the proton keeps moving and its direction of motion keeps changing under the influence of the force exerted upon it by the field. The fact that this force is always perpendicular to the direction of motion reminds us of circular motion - key features of which are that the net (centripetal) force is always directed perpendicularly to the circling object's motion (toward the center of the circle) and that the object's direction of motion is continually changing under the influence of this force (figure 10-17).

figure 10-17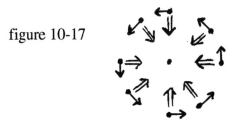

We conclude therefore that the path of the proton in the field is that of a circle whose plane is perpendicular to the field lines, and that the proton will be revolving in a counter-clockwise manner. The radius of the orbit can be determined by setting up an equality between the magnetic force exerted on the proton, $F = Bqv$, and the expression for the centripetal force acting on any object that undergoes circular motion, $F_c = mv^2/r$. The magnetic force *is* the centripetal force in this case.

LINK UNKNOWN TO GIVEN

$F = Bqv$ and $F_c = mv^2/r$

$Bqv = mv^2/r$

$Bq = mv/r$

$r = mv/Bq$

$r = (1.67 \times 10^{-27})(2 \times 10^6)/(1.6 \times 10^{-19})(5 \times 10^{-1})$

$r = .04$ meters

Magnetism

For this scenario to occur the field must be at least as large as the proposed circle. Otherwise, the proton exits the field at some point and proceeds to move from there in a straight line (figure 10-18).

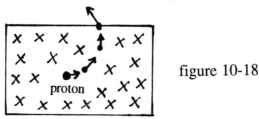

figure 10-18

(3) A straight wire segment is moved across a magnetic field, as illustrated in figure 10-19, with a velocity of 0.4 m/s. (a) In what

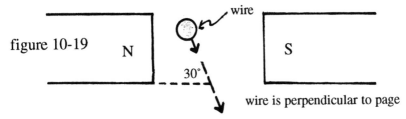

figure 10-19

direction will current be induced in the wire? (For the current to actually exist the wire segment must be part of a complete loop.) (b) What is the magnitude of the magnetic force exerted on each electron in the wire segment if the intensity of the field is 200 W/m^2 ?

SKETCH AND IDENTIFY

As the wire segment moves across the field, the electrons inside the wire are carried along with it. The electrons in the wire are therefore moving in a magnetic field, and magnetic fields exert forces on charges that move through them. (The protons in the wire are, of course, also carried across the field but, since it is extremely difficult to get them to move through the wire, we ignore them.) The direction of the magnetic field intensity between the poles is from left to right in figure 10-19. This is so since the direction assumed by the *S*-to-*N* arrow (S→N) of a test magnet placed in the field will be from left to right (like poles repel, unlike poles attract).

The velocity vector of the wire segment, **v**, can be resolved into two components, as illustrated in figure 10-20. Since a component of the velocity is perpendicular to the direction of the field intensity, the field exerts a force on the electrons in the wire. The direction of

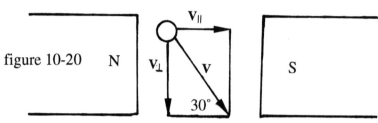

figure 10-20

this force can be determined by applying hand rule three. The magnitude of the force exerted on each electron can be ascertained from $F = Bqv$, keeping in mind that v in the formula now represents the component of the velocity perpendicular to the field.

LINK UNKNOWN TO GIVEN

(a) To find the direction of the force exerted by the field on the electrons in the wire, proceed as follows:

index finger - field intensity: →
thumb - component of motion perpendicular to field: ↓
middle finger (left hand) - force: XXX

Since the force acts parallel to the wire, it causes the electrons to flow through the wire. This *induced current* is directed perpendicularly to the page, away from the reader.

(b) To find the magnitude of the force exerted on each electron in the wire:

$F = Bqv$ (v here is $v_\perp = v \sin 30°$)

$F = (2 \times 10^2)(1.6 \times 10^{-19})(.4)(\sin 30°) = 6.4 \times 10^{-18}$ N

(4) A conducting wire in the shape of a rectangle is placed on the right side of an upward moving current of electrons, as illustrated in figure 10-21. The rectangle is 0.5 cm wide and 80 cm long, and the

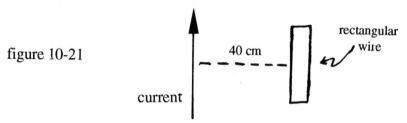

figure 10-21

near side of the rectangle is 40 cm from the current. (a) What is the direction of the magnetic flux lines passing through the plane of the rectangle? (b) How many flux lines are enclosed by the rectangle if the current is 3.2×10^{10} amperes? (Since the far end of the rectangle is only about one percent more distant from the current than the near end, this difference in distance may be ignored.)

SKETCH AND IDENTIFY

The direction of magnetic flux lines is everywhere the same as the direction of the magnetic field intensity and can therefore be determined via hand rule one. The magnitude of the field intensity can be found from the formula $B = \mu I/d$. This is equal to the density of field lines (*flux density*) expressed as the number of flux lines per square meter of area (perpendicular to the lines). Once the density per area is determined we can multiply that by the known area of the rectangle to find the total number of lines enclosed in the rectangle.

LINK UNKNOWN TO GIVEN

(a) According to hand rule one, the magnetic field intensity on the right side of the upward moving current of electrons in figure 10-22 is directed perpendicularly to the page, toward the reader.

This is therefore also the direction of the magnetic flux lines in the area enclosed by the rectangle.

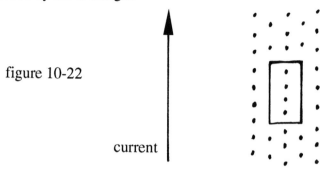

figure 10-22

current

(b) The magnitude of the magnetic field intensity in the area bounded by the rectangle can be determined as follows:

$B = \mu I/d = (2 \times 10^{-7})(3.2 \times 10^{10})/(4.025 \times 10^{-1})$

(40.25 cm, or .4025 meter, is the average distance from rectangle to current.)

$B = 1.59 \times 10^4$ W/m²

The density of field lines per square meter in the area enclosed by the rectangle is 1.59×10^4 W/m². The area of the rectangle is $(.8)(.005)$, or 4×10^{-3}, square meter. Therefore, the total number of field lines (customarily symbolized by the Greek letter ϕ) piercing the area enclosed by the rectangle is:

$\phi = BA = (1.59 \times 10^4)(4 \times 10^{-3}) = 63.6$ Webers

10.4 ON YOUR OWN PROBLEMS

(Asterisk indicates solution appears in appendices)

*(1) A wire carries a current of electrons perpendicularly to the page, away from the reader, as in figure 10-23. At each point in the figure, labeled A thru H, indicate the direction of the magnetic field intensity created by the current.

Magnetism

figure 10-23

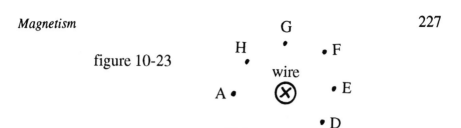

*(2) A current-carrying wire is situated between two magnetic poles, as in figure 10-24. The length of the wire in the field is 40 cm, the current is 15 amperes, and the magnitude of the field's intensity is 20,000 N/A·m. (a) In what direction will a force, if any, be exerted on the wire by the field? (b) How strong will the force be? (c) If the surface area of either pole is .04 square meter, how many flux lines run from pole to pole?

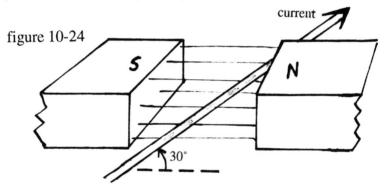

figure 10-24

*(3) What is the frequency of rotation of a proton in a magnetic field whose intensity is .06 W/m²? Prove that this frequency is independent of the velocity of the proton. (Assume the proton is moving perpendicularly to the field.)

*(4) An electron is projected horizontally into the region between two oppositely charged plates with a velocity of 4 x 10⁶ m/s, as in figure 10-25. The magnitude of the intensity of the *electric* field between the plates is 1000 N/C. What should be the direction and

magnitude of the intensity of a *magnetic* field in the region between the plates in order that the electron continues to move in a straight line?

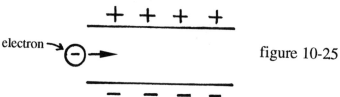

figure 10-25

*(5) Assuming you can see electrons and the activity they are engaged in, describe what you would see (insofar as the motion of electrons is concerned) as you look at the S-pole of a bar magnet (figure 10-26). In other words, what would be the predominant motion of the electons as seen by you from that vantage point?

figure 10-26

(6) Two currents of electrons, one of 6 amperes, the other of 3 amperes, are arranged parallel to each other. The wires carrying the currents are separated by 2 cm, the parallel arrangement is 5 meters long and the currents are oppositely directed. (a) Show that the field created by each current acts to repel the other current. (b) Determine the magnitude of the magnetic force exerted on each current.

(7) If the 3-ampere current in problem six is replaced by a single electron traveling in the same direction at the rate of 2×10^4 m/s, what is the magnitude and direction of the force exerted on the electron?

(8) An alpha particle (two protons, two neutrons) is projected into the magnetic field shown in figure 10-27 and proceeds to orbit

Magnetism

in the plane of the page under the influence of the magnetic force exerted upon it. The particle is projected with a velocity of 5 x 10^5 m/s and the intensity of the field is 20 W/m². (a) What is the radius

figure 10-27

Field

of the orbit? (b) Does the particle orbit in a clockwise or counter-clockwise manner? (c) What changes, if any, occur in the speed of the particle as it continues to circle in the field? (d) What is the period of rotation? (Assume the field is large enough to enclose the entire orbit, that every proton's and neutron's mass is 1.67 x 10^{-27} kg and that the charge on each proton is 1.6 x 10^{-19} C.)

(9) Describe the path of the alpha particle in problem eight if it enters the field perpendicularly to the page, away from the reader, with the same speed.

(10) A wire carrying 8 amperes of current is inserted into the magnetic field in problem eight (figure 10-27) such that the electrons in the wire flow from left to right. (a) What is the direction of the force exerted on the wire? (b) How strong is the force if the field is 50 cm long (and wide) and the wire stretches across the entire length of the field and beyond?

(11) Two opposite magnetic poles are arranged face-to-face. Each is 10 cm long and 4 cm wide, and the intensity of the field between them is a uniform 10,000 Tesla. How many magnetic flux lines exist in the region between the poles?

(12) A wire situated in the plane of this page carries a current of

electrons from right to left. Indicate the direction of the magnetic field intensity at each of the following locations: (a) directly above the wire, (b) directly below the wire, (c) in front of the wire and (d) behind the wire. Use the gollowing symbols : ↑,↓,→,←, x, •, and "none".

(13) A circularly shaped wire situated in the plane of this page carries a clockwise current of electrons. What is the direction of the magnetic field intensity (a) inside the circle and (b) outside the circle, in the plane of this page?

(14) A coil-shaped wire carries a current of electrons as illustrated in figure 10-28. Indicate the direction of the field intensity at each location in the figure labeled A thru F.

figure 10-28

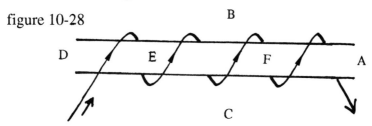

(15) How intense is the magnetic field at a point 4 cm distant from a 100-ampere current?

(16) A proton is projected into the magnetic field in figure 10-29 at an angle of 60° from the field. The proton enters the 10,000 Tesla field with a speed of 2×10^7 m/s. What is the direction and magnitude of the force exerted on the proton, as it enters the field?

figure 10-29

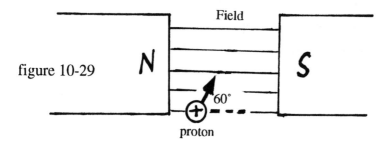

Magnetism

(17) What is the frequency of rotation of an electron moving perpendicularly to a magnetic field whose intensity is 200 W/m²? (The mass of an electron is 9.1 x 10⁻³¹ kg.)

(18) A proton is projected horizontally into the region between two oppositely charged plates with a velocity of 3 x 10⁵ m/s (figure 10-30). The intensity of the *electric* field between the plates is 500 N/C. What should be the direction and magnitude of the intensity of a *magnetic* field in the region between the plates in order that the proton continues to move in a straight line?

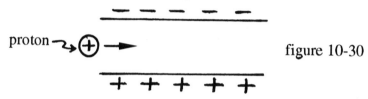

figure 10-30

(19) Two straight, parallel wires 6 meters long and 40 cm apart carry currents of 30 amperes in opposite directions. (a) What is the direction and magnitude of the net magnetic field intensity at a point midway between the wires? (b) What is the magnitude of the net field intensity at the same point if the currents are in the same direction?

(20) A wire carries an electron current from west to east in a uniform magnetic field that is directed from south to north. In what direction does the wire experience a magnetic force, if any?

(21) An alpha particle of mass 6.64 x 10⁻²⁷ kg moves in a circular path of radius 0.4 meter in a uniform magnetic field whose intensity is 4 W/m². Calculate (a) the speed of the particle and (b) the period of its revolution.

(22) Two long parallel wires carry current in opposite directions. The current in one wire is twice as large as the current in the other wire. Compare the force exerted by each wire on the other.

(23) A rectangular loop of conducting wire is placed in a horizontal magnetic field so that the plane of the loop is parallel to the field lines. A current is then made to flow around the loop. What will be the effect of the field upon the loop? (Hint: consider the force exerted on each side of the loop.)

(24) A horizontal length of wire four meters long weighs two newtons. When it is placed at right angles to a magnetic field and a current of 30 amperes is passed through it, the magnetic force just balances the wire's weight. How intense is the magnetic field?

(25) Three parallel current-carrying wires, each 5 meters long, pass through the corners of an equilateral triangle, as illustrated in figure 10-31. Currents A and C are directed out of the plane of the

figure 10-31

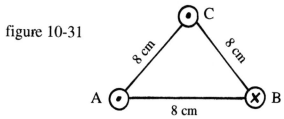

triangle (toward the reader); current B is directed into that plane (away from the reader). Each wire carries a current of 20 amperes; each side of the triangle is 8 cm long. What is the direction and magnitude of the net magnetic force exerted on wire C?

11

CHAPTER ELEVEN

MOMENTUM CONSERVATION

11.1 VOCABULARY AND FORMULAS

WORD	SYMBOL	UNIT
Impulse	I	Newton·second (N·s)
Momentum	p	Kilogram·meter/second (kg·m/s)

DEFINITIONS

IMPULSE - The net force exerted on an object multiplied by the time the force acts. Impulse is a vector quantity, oriented in the direction of the force.

MOMENTUM - The mass of an object times its velocity at any point in time. Momentum is a vector quantity, oriented in the direction of the object's velocity.

FORMULAS

a. $p = mv$ b. $I = Ft$ c. $Ft = \Delta mv$

d. $m_1v_1 + m_2v_2 = m_1v_1' + m_2v_2'$ e. $\mathbf{p}_T = \mathbf{p}_T'$

11.2 THINGS TO KNOW

(1) An object that moves to the right is customarily assigned a velocity and momentum value that is positive, and an object that moves to the left is assigned a velocity and momentum value that is negative. The same is true of objects that move upward (positive) vs. those that move downward (negative). All the equations listed above are valid only if motion in opposite directions are assigned velocity and momenta values that are opposite in sign.

(2) The principle of momentum conservation is applicable only to the *total* momentum of a *closed* system. A "closed" system is one in which all the interactions (forces) and all the actors (objects exerting the forces) associated with an event are included and taken into account. No part of a closed system may interact with anything that is not taken into account (and is therefore outside "the system").

As an example, consider the case of a rifle and a bullet. Before the trigger is pulled the total momentum of the system (rifle and bullet) is *zero* since no part of the system is moving. Pulling the trigger does change the momentum of the bullet (from zero to some positive number) and that of the rifle (from zero to some negative number, due to its recoil). But the total momentum of the closed system (the rifle and bullet) remains *zero* even after the trigger is pulled.

(3) The formula $m_1 v_1 + m_2 v_2 = m_1 v_1' + m_2 v_2'$ (Formulas, item *d*) is applicable only to one dimensional cases. Multi-dimensional problems are dealt with by resorting to the principle of *total momentum vector conservation* (Formulas, item *e*). That is, the total momentum vector after an event, $\mathbf{p_T}'$, is identical, in magnitude and direction, to the total momentum vector before the event, $\mathbf{p_T}$. Alter-

natively, the formula $m_1v_1 + m_2v_2 = m_1v_1' + m_2v_2'$ can be applied to each dimension independently of the other dimensions. It is then written as $m_1v_{1X} + m_2v_{2X} = m_1v_{1X}' + m_2v_{2X}'$ for the components of the velocities in the x dimension and $m_1v_{1Y} + m_2v_{2Y} = m_1v_{1Y}' + m_2v_{2Y}'$ for the components of the velocities in the y dimension (and so on for the z dimension in the case of a three dimensional problem). The x and y subscripts refer to the x and y components of the velocity vectors $\mathbf{v}_1$, $\mathbf{v}_2$, $\mathbf{v}_1'$ and $\mathbf{v}_2'$.

(4) Momentum conservation is a corollary of the third law of motion (see chapter five). Since forces always come in pairs that are equal in magnitude and opposite in direction, and these forces (the members of the pair) act simultaneously, their impulses (the product Ft) are also equal and opposite. Since impulse equals change in momentum (Formulas, item c), the momenta changes (of the objects upon which the paired forces act) are also equal and opposite. If one change is assigned a positive value (an increase) and the other a negative value (a decrease) there is no change in the total momentum of the system as a result of the event. This is automatically the case when the momenta of the objects are assigned opposite signs in opposite directions. If the change in total momentum is zero, then total momentum is conserved.

(5) The symbol F in formulas b and c represent the *net* force acting on an object.

11.3 SOLVING THE PROBLEMS

(1) A 2 kg mass moving to the right at 3 m/s collides head-on

with a 1 kg mass moving to the left at 4 m/s. After the collision, the 1 kg mass is observed to be moving to the right at 1 m/s. (a) How is the 2 kg mass moving after the collision? (b) What is the change in momentum of each mass as a result of the collision? (c) If the duration of the collision is 0.2 second, what average force was applied to each mass during the collision? (d) Are the forces and momenta changes equal in magnitude and with opposite sign?

SKETCH AND IDENTIFY

The phrase "head on" indicates that the centers of the objects were moving along the straight line drawn from one to the other, as illustrated in figure 11-1. This implies that neither mass goes off at

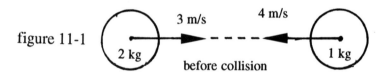

figure 11-1

before collision

an angle after the collision and that the problem, therefore, is a one dimensional one. The formula $m_1v_1 + m_2v_2 = m_1v_1' + m_2v_2'$ may therefore be used. Since the formula contains only one unknown,

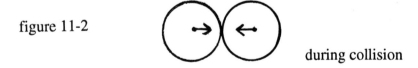

figure 11-2

during collision

namely v_1', use of the formula should reveal what the 2 kg object is doing after the collision. Then we can readily compare each object's initial momentum to its final momentum and ascertain its change in momentum. Since the change in momentum, Δmv, of each mass is equal to the impulse, Ft, that was applied to it during the collision (the event that produced the momentum change), and t is given, it

Momentum Conservation

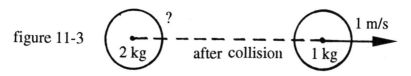
figure 11-3

should be a simple matter to calculate the average force, F, that was exerted on each mass. (The term *average* force is used here since the actual force may not have been constant throughout the collision time of 0.2 second. The formula $Ft = \Delta mv$ is then applicable to the average force.)

LINK UNKNOWN TO GIVEN

(a) $m_1v_1 + m_2v_2 = m_1v_1' + m_2v_2'$

$(2)(3) + (1)(-4) = 2v_1' + (1)(1)$

(V_2 is negative because object *two* was moving leftward before the collision.)

$2 = 2v_1' + 1$

$v_1' = .5$ m/s

Since v_1' is positive we conclude that the 2 kg mass continues to move to the right after the collision, at the rate of .5 m/s.

(b) To find Δmv for the 2 kg mass, proceed as follows:

mv before the event $= (2)(3) = +6$ kg·m/s

mv after the event $= (2)(.5) = +1$ kg·m/s

$\Delta mv = (+1) - (+6) = -5$ kg·m/s

Similarly, to find Δmv for the 1 kg mass:

mv before the event $= (1)(-4) = -4$ kg·m/s

mv after the event $= (1)(1) = +1$ kg·m/s

$\Delta mv = (+1) - (-4) = +5$ kg·m/s

(Note: the change in mv is always equal to the final mv minus the initial mv.)

(c) The force on the 2 kg mass is ascertained as follows:

$Ft = \Delta mv$

$(F)(.2) = -5$

$F = -25$ N

Analogously, to find the force on the 1 kg mass:

$Ft = \Delta mv$

$(F)(.2) = +5$

$F = +25$ N

(d) $F_1 = -25$ N and $F_2 = +25$ N

$\Delta(mv)_1 = -5$ kg·m/s and $\Delta(mv)_2 = +5$ kg·m/s

(2) A 0.5 kg baseball moving at 45 m/s is hit by a bat. The ball rebounds off the bat at the rate of 25 m/s. The collision time is estimated to be .05 seconds. How strong was the force that the bat exerted on the ball (on average)?

SKETCH AND IDENTIFY

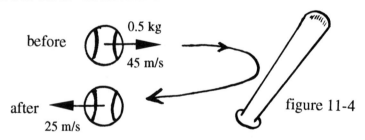

figure 11-4

The force exerted on the ball can be determined from $Ft = \Delta mv$ since we know t, mv before the event and mv after the event.

LINK UNKNOWN TO GIVEN

mv before the event: $(.5)(45) = +22.5$ kg·m/s

mv after the event: $(.5)(-25) = -12.5$ kg·m/s

$\Delta mv = (mv)_f - (mv)_i = (-12.5) - (+22.5) = -35$ kg·m/s

$Ft = \Delta mv$

Momentum Conservation

$$(F)(.05) = -35$$
$$F = -700 \text{ N}$$

The negative force value indicates that the force acted on the ball to the left. The ball exerted an equal and opposite force of +700 N to the right on the bat.

(3) A 600 gram billiard ball moving to the right at 2 m/s collides with an 800 gm ball at rest. After the collision the 600 gm ball takes off at 37° to the left of its original direction at the rate of 0.5 m/s, and the 800 gm ball takes off to the right of that direction, as illustrated in figure 11-5. What is the magnitude and direction of the 800 gm ball's velocity after the collision?

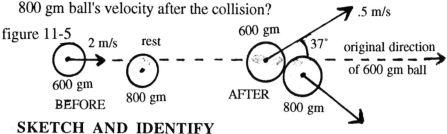

figure 11-5

SKETCH AND IDENTIFY

This problem can be solved either by applying the principle of total momentum vector conservation (method one) or by setting up equations based on the fact that the total momentum in each dimension must be independently conserved (method two).

method one

The total momentum vector before collision, $\mathbf{p}_T$, is found by determining the resultant of the momentum vectors of the two objects before collision. This is done in figure 11-6.

figure 11-6

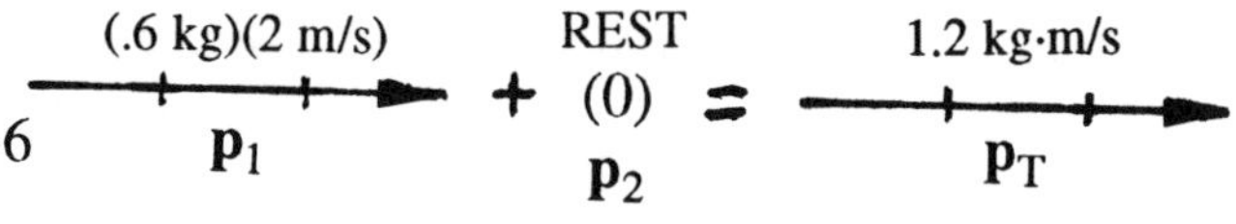

$$\underset{p_1}{(.6 \text{ kg})(2 \text{ m/s})} + \underset{p_2}{\text{REST} \atop (0)} = \underset{p_T}{1.2 \text{ kg·m/s}}$$

SCALE: 1 cm = .4 kg·m/s

The total momentum vector after the collision, p_T', is also determined by vector addition, the first step of which is illustrated in figure 11-7 (via the parallelogram method). The resultant of the two

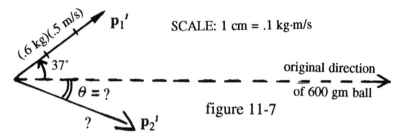

figure 11-7

momenta vectors after the collision, p_1' and p_2', must be identical to p_T, the total momentum vector before collision. That resultant is also the diagonal of the parallelogram formed with p_1' and p_2' as

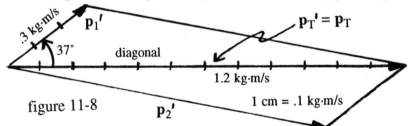

figure 11-8

adjacent sides (figure 11-8). From this the direction and magnitude of p_2' can be ascertained either by measurement (if the parallelogram is constructed to scale) or by calculation. From p_2' we can, in turn, find v_2', since $p = mv$.

method two

The principle of momentum conservation in the x dimension leads to the following equation:

$$m_1 v_{1x} + m_2 v_{2x} = m_1 v_{1x}' + m_2 v_{2x}'$$

The components v_{1x} and v_{2x} are given. The components v_{1x}' and v_{2x}' are found by vector resolution, as in figure 11-9.

Momentum Conservation

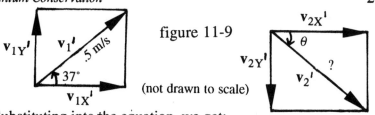

figure 11-9

(not drawn to scale)

Substituting into the equation, we get:

$(.6)(2) + (.8)(0) = (.6)(.5)(\cos 37°) + (.8)v_2' \cos \theta$

An analogous procedure for the y dimension yields:

$m_1 v_{1Y} + m_2 v_{2Y} = m_1 v_{1Y}' + m_2 v_{2Y}'$

$0 = (.6)(.5) \sin 37° + (.8)(-v_2' \sin \theta)$

From these two equations we can determine the two unknowns, v_2' and θ, the magnitude and direction of vector $\mathbf{v_2}'$.

LINK UNKNOWN TO GIVEN

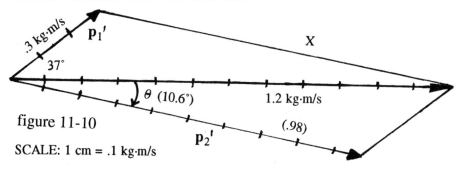

figure 11-10

SCALE: 1 cm = .1 kg·m/s

$x^2 = .3^2 + 1.2^2 - (2)(.3)(1.2)\cos 37°$ (Law of Cosines)

$x = .98$

The magnitude of $\mathbf{p_2}'$ is therefore equal to .98 kg·m/s.

$p_2' = mv_2'$ $.98 = (.8)v_2'$ $v_2' = 1.23$ m/s

$.3^2 = 1.2^2 + .98^2 - (2)(1.2)(.98)\cos \theta$

$-2.31 = -2.35 \cos \theta$

$\cos \theta = .983$

$\theta = 10.6°$

The direction of v_2' is 10.6° to the right of the 600 gm ball's original direction.

method two

In the x dimension: $1.2 = .24 + .8v_2' \cos \theta$

In the y dimension: $.18 = .8v_2' \sin \theta$

$v_2' \sin \theta = .225$ $\qquad v_2' \cos \theta = 1.2$

$v_2' \sin \theta / v_2' \cos \theta = .225/1.2 = \text{Tan } \theta = .187$

$\theta = 10.6°$

$v_2' \sin 10.6° = .225$

$v_2' (.184) = .225$

$v_2' = 1.22 \text{ m/s}$

(4) A 700 kg motorcycle moving at 20 m/s to the right collides head-on with a 1400 kg car moving to the left at 20 m/s. After the collision they remain locked together. What is the direction and magnitude of the motion of the combined mass after the collision?

SKETCH AND IDENTIFY

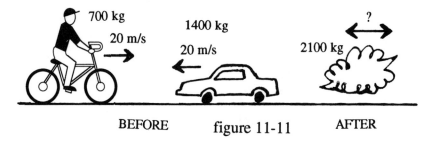

BEFORE   figure 11-11   AFTER

LINK UNKNOWN TO GIVEN

The total momentum before collision is $m_1v_1 + m_2v_2$. This is equal to $(700)(20) + (1400)(-20)$, or -14000 kg·m/s. After collision

Momentum Conservation

the total momentum is $m_T v'$, or $2100v'$. Since total momentum is conserved it must be true that:

$$2100v' = -14,000$$
$$v' = -6.67 \text{ m/s}$$

The combined mass of the car and motorcycle moves to the left at the rate of 6.67 m/s.

(5) A spaceship coasts at the rate of 40 m/s. In order to accelerate the ship its engine is turned on and two kilograms of oxidized fuel are exhausted per second from the rear nozzle of the ship. The particles of fuel emerge with a speed of 2,000 m/s in the direction opposite to the motion of the spaceship (figure 11-12). How much force does the engine exert on the spaceship?

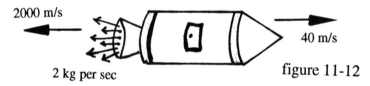

2000 m/s

2 kg per sec

40 m/s

figure 11-12

SKETCH AND IDENTIFY

Every second 2 kgs of fuel experience a reversal of direction. While inside the engine the fuel particles move with the spaceship, in the same direction and at the same rate (40 m/s). Once ejected, every 2 kg mass of fuel moves in the opposite direction at the rate of 2,000 m/s. Thus, every second two kilograms of matter experience a change in momentum from (2)(40), or +80 kg·m/s, to (2)(-2000), or -4000 kg·m/s. This provides us with enough information to apply the formula $Ft = \Delta mv$ and thereby ascertain the magnitude of the force exerted on the two kilograms of fuel. Since forces come in equal and opposite pairs this must also be the magnitude of the force exerted on the ship (immediately after the engine is turned on).

LINK UNKNOWN TO GIVEN

$$Ft = \Delta mv$$
$$(F)(1) = (-4000) - (+80)$$
$$F = -4080 \text{ N}$$

The force on the spaceship therefore is +4080 N.

11.4 ON YOUR OWN PROBLEMS

(Asterisk indicates solution appears in appendices.)

*(1) A 200 N skater standing on ice throws a 10 N ball to the right. The ball emerges from her hand with a velocity of 3 m/s. Assuming no friction between skater and ice, what is the magnitude and direction of the skater's velocity after the ball is thrown?

*(2) A 50 gram bullet is aimed at a 5 kg block of wood at rest. The bullet penetrates the block and remains embedded in it as the block takes off with a velocity of 10 m/s. What was the speed of the bullet before it encountered the block?

*(3) An open cart moves over a horizontal frictionless surface at the rate of 20 m/s. The mass of the cart is 400 kg. It then starts to rain and 2 kg of water enter (and remain in) the cart per second. What is the velocity of the cart two minutes after it started to rain? (Assume the raindrops fall vertically into the cart.)

*(4) In the game of baseball a well-trained catcher allows his hand to move backward every time a pitched ball is to be brought to a stop in his glove. (a) Why is this advantageous to the catcher? (b) If a 2 kg ball enters his glove at the rate of 40 m/s, and the catcher draws his hand back so that the ball comes to rest within a distance of 50 centimeters, how much force did the ball exert on his hand?

*(5) A billiard ball moving to the right at 5 m/s collides with

Momentum Conservation 245

another billiard ball of equal mass moving to the left at 2 m/s. As a result of the collision the first ball makes a 60° turn to the right of its original direction and the second ball makes a 150° turn to the right of *its* original direction. What is the magnitude of the velocity of each ball after the collision?

(6) A billiard ball of unknown mass is moving at 4 m/s. This ball overtakes a second ball of mass 15 kg which is moving in the same direction at 1 m/s. After the collision, the first ball reverses course and travels at 2 m/s in the opposite direction. The second ball is now moving at 5 m/s. What is the mass of the first ball?

(7) What is the change in momentum of each ball in problem six as a result of the collision? Indicate whether each change is positive or negative.

(8) If the duration of the collision in problem six is .08 second, how strong was the force of impact on each ball?

(9) A 20 kg projectile leaves a 1000 kg launcher with a velocity of 80 m/s. What is the recoil velocity of the launcher?

(10) A 2 kg baseball moving at 25 m/s rebounds off a bat at the rate of 35 m/s. The interaction time between bat and ball is 0.2 seconds. How strong was the force exerted by the *ball* on the *bat*?

(11) A 2 kg sphere moving at the rate of 8 m/s collides with a second stationary sphere, also of mass 2 kg. After the collision, the first ball takes off in a direction 60° to the left of its original direction. The second ball moves off in a direction 30° to the right of the first ball's original direction. What is the speed of each ball after the collision?

(12) A compressed spring is placed between two carts which are held together by a string. The string is then cut and the carts move apart. One cart, of mass 3 kg, takes off to the right with a velocity of

15 m/s. What is the velocity of the other cart whose mass is 2 kg? Assume no friction exists to slow the carts as they move apart.

(13) The engine of a rocket of unknown mass is fired for 40 seconds. During this time 100 kg of oxidized fuel is exhausted through the rear nozzle of the rocket. The particles of fuel emerge from the nozzle with an average velocity of 3000 m/s. The rocket, which used to be at rest, is now moving at the rate of 80 m/s. (a) What is the mass of the rocket at the end of the forty second period? (b) What average force was exerted on the rocket?

(14) A 3000 kg car leaves a parking lot. One minute later it is moving on a highway at the rate of 8 m/s. The coefficient of kinetic friction between the tires and the ground is (0.2). What average force did the engine exert on the car?

(15) A girl weighing 150 N and running at the rate of 3 m/s jumps onto a stationary sled of mass 10 kg. What is the speed of girl and sled on the level, frictionless ice?

(16) How much of an impulse did the girl in problem fifteen apply to the sled? How much of an impulse did the sled impart to the girl?

(17) What is the velocity of the cart in problem three if instead of water coming into the cart, water leakes out at the same rate for two minutes?

(18) A 32,000 kg airplane is moving horizontally at the rate of 20 m/s. It then fires a projectile of mass 200 kg in the forward direction. The muzzle velocity of the projectile is 150 m/s. What is the direction and magnitude of the airplane's velocity after the projectile is launched?

(19) A body of mass m moving with velocity v collides into a stationary body that is three times as massive. The two bodies then

Momentum Conservation

proceed to move together in the same direction. Compare the common velocity of the two bodies after collision to the original velocity of mass m before collision.

(20) Two opposing forces, one of 20 N, the other of 50 N, act on the same object for 12 seconds. What is the change in momentum of the object?

12

CHAPTER TWELVE

FORMS OF ENERGY

12.1 VOCABULARY AND FORMULAS

WORD	SYMBOL	UNIT
Work	W	Joule, J (N·m)
Power	P	Watt, W (J/s)
Energy	E	Joule or kwhr
Kinetic Energy	KE	Joule
Potential Energy	PE	Joule
Spring Constant	k	N/m
Potential difference	V	Volt, V (J/C)
Heat	Q	Joule or calorie
Specific Heat	c	J/gm·°C
Current	I	Ampere, A (C/s)
Resistance	R	Ohm, Ω (V/A)

DEFINITIONS

(1) WORK - Work is done by a force that is exerted on an object in the direction of the object's motion. If the object is not moving,

Forms of Energy 249

or if the force is perpendicular to the object's motion, no work is done. The amount of work done is provided by the formula $W = Fd\cos\theta$ where θ is the angle between the force F and the distance d traveled by the object. One newton·meter of work is referred to as one *joule*. When the angle between F and d is 0°, as is frequently the case, the cosine of θ is *one* and the formula becomes $W = Fd$.

(2) POWER - The rate of doing work. A rate of one joule per second is referred to as one *watt* of power.

(3) ENERGY - The ability to do work, in joules.

(4) KINETIC ENERGY - The ability to do work due to an object's motion.

(5) POTENTIAL ENERGY - The ability to do work due to position, arrangement or circumstances.

(6) SPRING CONSTANT - The ratio of the force of tension exerted by a stretched or compressed spring, in newtons, to the amount of stretch or compression, in meters.

(7) POTENTIAL DIFFERENCE - The amount of work that can be done by an electric field on one coulomb of charge as the charge is moved from one point in the field to another. A potential difference (p.d.) of one volt represents one joule of work per coulomb of charge. The potential difference between two points is sometimes referred to as the *voltage* between the points.

(8) INTERNAL ENERGY - The sum of the energy of all the molecules of an object, including all forms of energy, but not counting energy the object has as a whole.

(9) HEAT - Internal energy that is transferred from one object to another or from one part of an object to another.

(10) SPECIFIC HEAT - The amount of energy that must be added to or extracted from one gram of an object to change its tem-

perature (up or down) by one degree centigrade (°C). This quantity generally varies from material to material.

(11) CURRENT (of charge) - The rate of flow of charge past any particular point in a material. A flow of one coulomb per second is labeled one *ampere*. The symbol for current is I.

(12) RESISTANCE (electrical) - The slope of the voltage vs. current graph for a particular material. The resistance of a material through which one ampere flows when the potential difference between its ends is one volt is referred to as one *ohm*.

FORMULAS
a. $W = Fd\cos\theta$ b. $P = W/t$ c. $KE = mv^2/2$ d. $PE = kx^2/2$
e. $PE = mgh$ f. $W = Vq$ g. $V = Ed$ h. $Q = cm\Delta T$ i. $P = VI$
j. $V = IR$ k. $E = I^2Rt$ l. $W_{NET} = \Delta KE$ m. $E = pt$

12.2 THINGS TO KNOW

(1) Be careful with the use of such words as *work* and *power*, whose physics meaning is different from the way the words are employed in ordinary conversation. *Work* in physics is not synonomous with "making an effort", nor is it an activity that makes you tired or that you get paid for. If you stand in place and hold up a one-thousand pound weight until beads of sweat emerge from every pore in your skin, you are still not doing an iota of work (on the weight) as far as physics is concerned. Work in physics is done only when the quantity $Fd\cos\theta$ is greater than zero and the conditions described earlier (Definitions, item *one*) are fulfilled.

By the same token, *power* in physics is not akin to "strength" or

Forms of Energy 251

to "ability to do work". The former is actually synonomous with the physics term *force* and the latter is referred to in physics as *energy*. Power, as far as physics is concerned, is the rate of work actually done per time.

(2) The symbol F in the formula $W = Fd\cos\theta$ does not represent net force, as it does in $F = ma$. Any force that acts on an object in the direction of the object's motion, whether it is the only force acting on the object or merely one of many, does an amount of work on the object equal to $Fd\cos\theta$, where F is the magnitude of the force. Work done by a net force, however, is referred to as *net work* and is sybolized by W_{NET} (Formulas, item *l*).

(3) The short-cut formula for work, $W = Fd$, can only be used for forces or components of forces that act parallel to, and in the same direction as, an object's motion ($\cos\theta = \cos 0° = 1$). If this is not meticulously adhered to, incorrect conclusions will be drawn without any warning given to alert you to the fact that the result is wrong.

(4) Having energy is not the same as, nor does it mean that, work is being done, is about to be done or will ever be done. Instead, the existence of energy implies that work *could be* done but has not yet been done. As a matter of fact, once the work is done the ability to do that work (in other words - the energy) no longer exists with the object that had it, in the form that it used to exist.

For example, a moving object is endowed with kinetic energy in the sense that so long as it is moving it has the ability to use its motion to do work. It *may* collide with another object and, in the process, do work on that object. Should such a collision occur, however, the moving object will lose some or all of its speed and thereafter will no longer have the ability to do work based on its motion. If it loses only some of its speed, its ability to do work is reduced by an amount equal to the work performed. Should it, on the other hand, never collide with another object, it will retain all its ability to do work, but will never actually do the work it is capable of doing.

It's like "having the cake and eating it." As the saying goes, "you cannot have the cake and eat it too." If you have the cake, you haven't eaten it; if you have eaten it, you no longer have it. Analogously, you cannot have the energy and do the work. If you have the energy, you haven't done the work; if you have done the work, you no longer have the energy. (The energy does not disappear entirely, however. Some other body picks it up or it changes form or both.)

(5) The formulas for potential energy can only be used with a chosen *base level* in mind - a point or line at which the potential energy ia assigned a value of *zero*. For example, the formula for gravitational potential energy, $PE = mgh$, can only be used if some line is chosen from which all h (height) values are to be measured. That line has a h value of zero and the PE there is zero; The h value at any other point is the distance from the point to the chosen line.

Base levels should be chosen wisely in order to simplify the task of solving a problem. The base level in the case of a pendulum problem, for example, should conveniently be chosen to be the horizontal line running through the lowest point of the pendulum's swing. In the case of an object rising and falling inside a house, the base level might be the floor. If all the activity occurs over a table, however, it is probably best to make the surface of the table the base level.

(6) Potential energy exists only in the case of *conservative* forces

Forms of Energy 253

- such as gravity and electricity - for which the work done between two points is independent of the path taken between the points. For non-conservative forces - such as friction - no potential energy formula can be written or designed, since the work done by such forces does depend on the path taken by the object - something that is not under the exclusive control of the force.

(7) Although the symbol E throughout this chapter represents energy, in the formula $V = Ed$ it represents electric field intensity (see chapter nine). The formula is applicable only to electric fields of uniform intensity, such as that between two oppositely charged parallel plates. In the formula $W = Vq$, q is the charge acted upon by the electric field (the charge upon which the work is performed), not the charge that sets up the field. Don't confuse this W and others that appear in formulas to represent *work* with the W that represents the unit of power, the *watt*.

(8) The terms *potential difference* and *voltage* should be used only in reference to two points. We speak of the p.d. or voltage "between points A and B", such as between the terminals of a battery. Or we might say, "the p.d. *across* resistor X is such and such", referring to the p.d. between the ends of the resistor. Never apply the term *voltage* without identifying the two points between which the voltage exists. There is no such thing as voltage *at* point X; there is only voltage *between* points X and Y.

12.3 SOLVING THE PROBLEMS

(1) A 100 kg mountain climber lifts herself up a vertical distance of 20 meters, per minute. What is the power of this activity in watts,

kilowatts and horsepower? (1 kw = 1000 watts; 1 h.p.= 750 watts.)

SKETCH AND IDENTIFY

Assuming the climbing activity occurs at a uniform rate (the climber does not accelerate or decelerate), the net force on the climber must be *zero*. This means that the upward force exerted on her (in reaction to the downward force she exerts on the rope or on protruding pieces of the mountain) is just equal, on average, to her weight. Since the climber's weight can be ascertained from $W = mg$, the magnitude of the upward force acting on her is known. This upward force, called the *normal*, acts in the direction of the climber's motion and therefore does work on her. The angle between the normal and the climber's motion is zero, since both are directed vertically upward (figure 12-1). The work done per minute

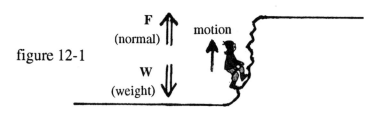

figure 12-1

can therefore be determined from $W = Fd\cos\theta$, since F (the normal), d (the distance traveled per minute) and θ are now all known. Dividing this amount by sixty then yields the work done per second and the power in watts.

LINK UNKNOWN TO GIVEN

$W = mg = (100)(9.8) = 980$ N

$F = W = 980$ N

$W = Fd\cos\theta$

$W = (980)(20)(1) = 19{,}600$ N·m   (per minute)

Forms of Energy

Work per second = 19,600/60 = 326.67 J/s

The power is 326.67 watts.

The power in kw = watts/1000 = 326.67/1000 = .33 kw

The power in h.p. = watts/750 = 326.67/750 = .44 h.p.

(2) What is the average power of an engine that accelerates a 10,000 kg car from 30 m/s to 50 m/s in 5 seconds? Assume no friction and express your answer in horsepower (h.p.).

SKETCH AND IDENTIFY

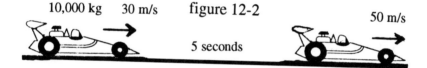

From the change in speed (Δv) per time (Δt) the acceleration rate can be determined. From $F_{NET} = ma$ and the fact that the only force exerted on the car is that provided by the engine (there is no friction) we can compute the magnitude of the engine force. That force acts in the direction of the car's motion ($\cos\theta = \cos 0° = 1$) and therefore does work. The distance traveled is provided by the formula (see chapter two) $d = v_i t + at^2/2$. Multiplying the distance by the force yields the amount of work done. That number is then divided by the time to obtain the power.

LINK UNKNOWN TO GIVEN

$a = \Delta v / \Delta t =$ (50-30)/5 = 20/5 = 4 m/s²

$F_{engine} = F_{NET} = ma =$ (10,000)(4) = 40,000 N

$d = v_i t + at^2/2$

$d =$ (30)(5)+ (4)(5)²/2 = 200 m   (five-second interval)

$W = Fd\cos\theta$

$W = (40{,}000)(200)(1) = 8 \times 10^6$ J   (five-second interval)

$P = W/t = (8 \times 10^6)/5 = 1.6 \times 10^6$ W

Converting to horsepower:

$(1.6 \times 10^6)/750 = 2.13 \times 10^3$ h.p.   (1 h.p. = 750 watts)

(3) A 200 kg wrecking ball is dropped from a height of 20 meters onto a nail waiting to be driven into the ground. After falling freely through the twenty-meter distance, the ball lands on the nail and proceeds to push it into the ground a distance of 10 cm. (a) What was the *PE* of the ball before it was dropped? (b) What is the *KE* of the ball as it collides with the nail? (c) What average force does the ball exert on the nail? (d) How much work does the ball do on the nail?

SKETCH AND IDENTIFY

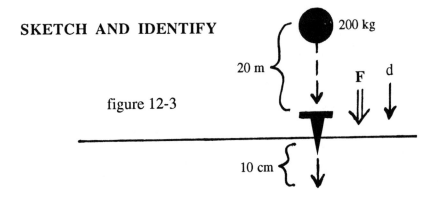

figure 12-3

The gravitational *PE* of the ball at the 20 meter position can readily be ascertained from the equation $PE = mgh$. Its velocity at the end of the twenty meter trip - just as it strikes the nail - can be computed from $v_f^2 - v_i^2 = 2ad$ (see chapter two) with *a* equal to 9.8 m/s², the acceleration of free fall. The *KE* of the ball as it collides with the nail can then be determined from the equation $KE = mv^2/2$.

Forms of Energy

Since the nail is driven a distance of 10 cm into the ground, we know that the ball is brought to a stop in that distance. We therefore know the speed of the ball at the beginning and at the end of its 10 cm push into the ground. The formula $v_f^2 - v_i^2 = 2ad$ is now applied once again, this time to determine the deceleration rate of the ball as it pushes the nail into the ground. With the help of $F_{NET} = ma$ we next calculate the average force acting on the ball (exerted by the nail) to slow it down and ultimately stop it. This force is equal and opposite (action - reaction) to that exerted on the nail (by the ball) to drive it into the ground. The formula $W = Fd\cos\theta$ then yields the amount of work done by the ball on the nail.

LINK UNKNOWN TO GIVEN

(a) $PE = mgh$ (before release)

$PE = (200)(9.8)(20) = 39,200$ J

(b) $v_f^2 - v_i^2 = 2ad$ (20 meter trip, free fall)

$v_f^2 - 0 = 2(9.8)(20)$

$v_f = 19.8$ m/s

The ball arrives at the nail with a velocity of 19.8 m/s.

$KE = mv^2/2$ (at end of 20 meter trip)

$KE = (200)(19.8)^2/2$

$KE = 39,200$ J

(The fact that the *KE* at the bottom is the same as the *PE* on top is no accident. It indicates that the lost *PE* is turned into *KE* with no net change in the energy of the ball.)

(c) $v_f^2 - v_i^2 = 2ad$ (ball's 10 cm push into nail)

$0 - 19.8^2 = (2)(a)(.1)$ (10 cm = .1 m)

$a = -1960$ m/s²   (average deceleration of the ball)

$F_{NET} = ma = (200)(1960)$

$F_{NET} = 3.92 \times 10^5$ N

The average force exerted on the ball by the nail and on the nail by the ball is 392,000 N. (The weight of the ball is relatively insignificant compared to a force of this magnitude, so we are justified in ignoring it when analyzing the nail-ball interaction.)

(d) $W = Fd\cos\theta$ (work done by ball on nail)
    $W = (3.92 \times 10^5)(.1)(1)$ (F and d in same direction)
    $W = 39,200$ J

The ball does all the work it was capable of doing as a result of its velocity (*KE*) and, before that, its position (*PE*).

(4) A 3 Ω conducting wire is connected to the terminals of a 9-volt battery, as illustrated in figure 12-4. (a) How much work does the battery do on every electron that makes the trip from the negative terminal to the positive terminal? (b) What is the current in the circuit? (c) What is the power of the circuit? (d) How much work does the battery do, and how much energy does it lose, in 15 minutes of operation? (e) Assuming all the work is done internally on the molecules of the wire - no heat loss to the surrounding air - by how many °C will the temperature of the wire increase if its mass is 1,000 grams and its specific heat is .39 J/gm·°C?

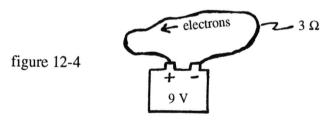

figure 12-4

SKETCH AND IDENTIFY

The voltage of a battery is, by definition, the work it does on one coulomb of charge as the coulomb goes from terminal to

Forms of Energy

terminal (usually a coulomb of negatively charged electrons going from the negative terminal to the positive terminal). The work done by the battery on some other amount of charge, such as one electron, is determined from the equation $W = Vq$, where q is the amount of charge on which the work is done. The shape or length of the wire connecting the terminals is immaterial to the amount of work done, since electricity is a *conservative* force (Things to Know, item 6) - the work done is independent of the path taken between two points (the two terminals in this problem).

The current, in amperes, can be determined from $V = IR$. The power, in watts, can then be found from $P = VI$ and the work done over a period of time (such as 15 minutes) can then be ascertained from $P = W/t$.

The work done by a *net* force acting on an object is equal to the change in the object's kinetic energy (Formulas, item *l*). Thus, if all the work done by the battery is applied (via the action of the electrons) to the molecules of the wire, the wire's internal *KE* - the *KE* of its molecules - will increase by an amount equal to the work done. The wire's gain in internal *KE* is, in turn, reflected in its temperature via the relationship $Q = cm\Delta T$.

LINK UNKNOWN TO GIVEN

(a) $W = Vq$

$W = (9)(1.6 \times 10^{-19})$ (q is charge on one electron)

$W = 1.44 \times 10^{-18}$ J

(b) $V = IR$

$9 = I(3)$

$I = 3$ A

The rate of flow of charge is 3 coulombs per second or 3 amperes.

(c) $P = VI$

$P = (9)(3) = 27$ W

The rate of doing work (the power) is 27 joules per second or 27 watts.

(d) $P = W/t$

$27 = W/900$ (15 minutes is 900 seconds)

$W = 24{,}300$ J

The battery does 24,300 joules of work and, therefore, loses 24,300 joules of energy in 15 minutes.

(e) $Q = cm\Delta T$

$24{,}300 = (.39)(1000)(\Delta T)$ (units: joules, grams, degrees)

$\Delta T = 62.3\ °C$

(5) A pump raises 300 kg of water per minute from a well 20 meters deep. The water emerges from the well with a velocity of 30 m/s. What is the power of the pump? Assume no friction.

SKETCH AND IDENTIFY

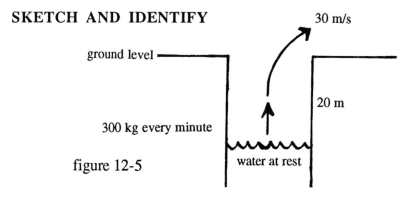

figure 12-5

The force exerted by the pump must be greater than the weight of the water carried upward, since the water is accelerated from rest to 30 m/s (figure 12-6). From $v_f^2 - v_i^2 = 2ad$ we can calculate the upward acceleration rate of the water and from $F_{NET} = ma$ the net force exerted on the water can be determined. This net force is equal

Forms of Energy

to the difference between the upward force exerted by the pump and the downward force exerted by the weight of the water. Since the weight can readily be determined from $W = mg$, it should not be difficult to ascertain the upward force exerted by the pump. The equation $W = Fd\cos\theta$ then yields the work done, every minute, by the pump and $P = W/t$ provides the power of the pump.

figure 12-6

$F_p > W$ | water 300 kg | $F_{NET} = F_p - W$

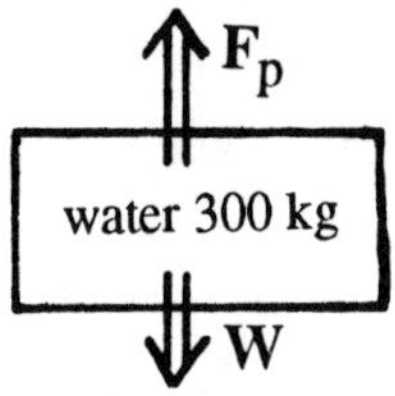

LINK UNKNOWN TO GIVEN

$v_f^2 - v_i^2 = 2ad$

$30^2 - 0 = 2(a)(20)$ (upward trip of water)

$a = 22.5 \text{ m/s}^2$

$F_{NET} = ma$

$F_p - W = ma$ (F_p is force exerted by *pump*)

$F_p - mg = ma$

$F_p = m(a + g)$

$F_p = (300)(22.5 + 9.8) = 9{,}690 \text{ N}$

$W = F_p d \cos\theta$

$W = (9690)(200)(1)$ ($\theta = 0°$)

$W = 193{,}800 \text{ J}$ (work done per minute)

$P = W/t = 193{,}800/60 = 3230 \text{ J/s} = 3230 \text{ W}$

ALTERNATE METHOD

The force exerted by the pump consists of two components (not necessarily equal to each other, but in the same direction). One component acts to neutralize ("cancel out") the downward acting pull of gravity. If this were all the force exerted by the pump, the net force on the water would be *zero* and the water would not accelerate. Instead, the pump's force exceeds the water's weight. This excess force is the second component. It constitutes the net force that causes the water to accelerate.

The work done by that component of the pump's force that neutralizes the water's weight is referred to as the *work done against gravity*. The work done by the excess force - the net force - is referred to as the *net work* (Things to Know, item two). The net work is equal to the change in *KE* it produces in the water it acts upon (Formulas, item *l*).

The total work done by the pump, W_p, is equal to the sum of the work done by its component forces - the work done against gravity, W_g, plus the net work, W_N.

$W_p = W_g + W_N$

$W_g = F_g d \cos\theta$

$W_g = mgd\cos\theta = (300)(9.8)(20)(1) = 58{,}800$ J

$W_N = \Delta KE = mv_f^2/2 - mv_i^2/2$

$W_N = (300)(30)^2/2 - 0 = 135{,}000$ J

$W_p = 58{,}800 + 135{,}000 = 193{,}800$ J

This agrees with our previous conclusion regarding the work done by the pump every minute.

(6) A 5 kg mass hangs at the end of a stretched spring. The spring constant is 100 N/m. How much elastic *PE* is stored in the spring?

Forms of Energy

SKETCH AND IDENTIFY

Since the 5 kg mass "hangs" motionless, it must be in a state of equilibrium. This means that the force of tension exerted upward by the stretched spring, F_s, is just equal to the weight, W, of the 5 kg mass acting downward (figure 12-7). Knowledge of the weight (from $W = mg$) is therefore tantamount to knowing the force of tension. From the fact that the spring exerts a tension force of 100 N for every meter of elongation and Hooke's law for springs ($F_s = kx$) we can compute the elongation, x. The formula $PE = kx^2/2$ then reveals the amount of elastic PE stored in the spring.

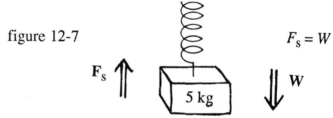

figure 12-7

LINK UNKNOWN TO GIVEN

$W = mg = (5)(9.8) = 49$ N

$F_s = W = 49$ N

$F_s = kx$

$49 = 100(x)$ (k is spring constant in N/m)

$x = .49$ m

$PE = kx^2/2 = (100)(.49)^2/2 = 12$ J

12.4 ON YOUR OWN PROBLEMS

(Asterisk indicates solution appears in appendices.)

*(1) A machine operating at 500 watts of power pulls on a cable that drags a heavy box along the floor. The cable makes an angle of 30° with the horizontal and the box moves at the constant rate of five

m/s. How strong is the force exerted on the box?

*(2) A proton is situated in the region between two oppositely charged plates separated by 2 cm. (a) How much work is done on the proton as it is pushed from the positive to the negative plate if the electric field intensity between the plates is 1000 N/C? (b) What is the *KE* of the proton as it arrives at the negative plate, assuming it started from rest?

*(3) What does it cost to operate a 240 Ω appliance for 5 hours at the rate of 15 cents per kwhr if the utility company provides all outlets with a p.d. of 120 volts?

*(4) A cup of tea at 100 °C is allowed to cool to room temperature, 30 °C. The cup contains 100 grams of liquid which consists almost entirely of water (specific heat: 1.0 cal/gm·°C). If all the internal energy lost by the tea were put to work lifting a one kilogram object vertically upward, how high would the object rise? One calorie is 4.18 joules.

*(5) A 10 kg block slides down an 80-meter long incline from a height of 16 m, as illustrated (figure 12-8). The coefficient of kinetic friction (see chapter six) between the block and the incline is (.1). (a) How much work does gravity do against friction as the block slides down the length of the incline? (b) Assuming the block starts from rest, what is the *KE* of the block as it arrives at the bottom of the incline?

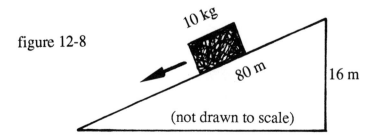

figure 12-8

Forms of Energy

(6) A weight lifter picks up a 60 kg dumb-bell and raises it to a height of 3 meters in 4 seconds. What is the average power of this activity?

(7) The weight lifter in problem six now holds the dumb-bell in place, at a distance of 3 meters from the floor, for 2 minutes. What is the power of this activity?

(8) A spaceship in a stable circular orbit around and near earth completes one revolution every 90 minutes. (a) How much work does the earth's gravitational pull do on the ship per revolution? (b) What is the power of this activity? (Careful!)

(9) A car engine operating at its full power rating of 150 h.p. acts to keep a 6000 N car moving at constant speed. The coefficient of rolling friction between the tires and the road is (.2). What is the speed of the car?

(10) A 4 kg block moving at 20 m/s collides with a 10 kg block at rest on a frictionless surface. When the 10 kg block has been pushed a distance of 2 meters, the 4 kg block has lost all its speed and comes to rest. (a) What is the average deceleration rate of the 4 kg block? (b) How strong is the average force exerted on the 10 kg block by the collision? (c) How much work did the 4 kg block do on the 10 kg block? (d) Compare your answer to part *c* with the *KE* of the 4 kg block just before the collision. What is the significance of this comparison?

(11) A 60 watt lamp connected to a 120 volt outlet operates for 3 hours. (a) What is the current through the lamp? (b) How much work, in joules and kwhrs, does the source (the utility's generator) do during the three hours of operation? (c) How much energy in the form of light and heat is provided by the circuit during the three hours? (d) How many coulombs of charge pass through the circuit

during the three hours? (e) What is the resistance of the circuit?

(12) A 4,000 kg elevator at rest is accelerated vertically upward at the rate of 2 m/s². The motor responsible for this acceleration acts to raise the elevator a distance of 40 meters against a force of friction of 8,000 N. (a) How much work does the motor do? (b) What is the average power of this activity? (c) What is the *net* work done on the elevator? (d) What is the *KE* of the elevator at the 40 meter mark?

(13) A spring whose constant is 12 N/m is compressed a distance of 4 cm. (a) How strong is the force exerted by the spring in this condition? (b) How much work can the spring do?

(14) You push into the handle of a lawnmower with a force of 200 N. The handle makes an angle of 60° with the horizontal, and the lawnmower moves forward at the constant rate of 0.3 m/s. What is the power of this activity in watts and h.p.?

(15) A pith ball carrying a negative charge of 6 x 10^6 C is situated in the region between two oppositely charged plates separated by 5 mm. As the pith ball is pushed from the negative to the positive plate, 100 joules of work are done on it by the electric field. What is the intensity of the electric field between the plates?

(16) How many joules of work are equivalent to one kwhr? At a cost of 15 cents per kwhr, charged by some utility companies, what is the cost of one joule?

(17) A 12 kg object is held at a distance of 8 meters from the ground. The object is then dropped and allowed to fall freely until it crashes into the ground below. (a) What was the *PE* of the object before it was dropped? (b) How much work did gravity do on the object as it fell to the ground? (c) Calculate the speed of the object just before it collides with the ground. (d) What is the *KE* of the object as it strikes the ground? (e) Compare your answers to parts *b*

Forms of Energy

and *d*. What is the significance of this comparison?

(18) A frictionless incline is oriented θ degrees from a horizontal floor. The incline is *d* meters long and rises to a height of *h* meters from the floor. Show that the work done in lifting an object of mass *m* vertically upward to the top of the incline is the same as the work done in pushing the object up the incline, assuming both motions occur at constant speed.

(19) One thousand kilograms of water fall freely over a 200 meter water-fall per minute. (a) How much work can this falling water do every minute in driving the turbine of a generator situated at the bottom of the fall? (b) If all this work is used to run an electric current through a circuit, what is the power of the circuit?

(20) A carpenter uses a 2 kg sledge hammer to drive a nail into concrete. Each blow drives the nail 1 cm deeper into the concrete. If the carpenter raises the hammer 60 cm each time he prepares to strike the nail and allows the hammer to fall onto the nail under the influence of its own weight, what is the average force exerted on the nail during each blow?

(21) A 100 gm ball is placed in front of a compressed horizontal spring whose constant is 4,000 N/m. The spring is compressed a distance of 8 cm. What is the maximum velocity of the ball after the spring is released?

(22) The spring and ball of problem twenty-one are now placed inside the barrel of a gun in which the ball encounters a 50 N strong resistive force of friction. Assuming the spring, when released, reaches all the way across the barrel, what is the maximum velocity of the ball after the spring is released?

13

CHAPTER THIRTEEN

CONSERVATION OF ENERGY

13.1 VOCABULARY AND FORMULAS

DEFINITIONS

ENERGY CONSERVATION - The total energy of a system, taking all forms of energy into account, remains unchanged. A *system* is a set of objects none of which experience an external force (a force exerted from outside the system).

MECHANICAL ENERGY - The sum of the kinetic energy (KE) and potential energy (PE) of a system.

LENZ'S LAW - All magnetic effects lead to forces that oppose the change that creates the effect. In other words, all magnetic effects self destruct unless the change that creates the effect is maintained externally (by an agent outside the change-effect-force loop).

BACK *EMF* - The force that opposes the change that creates the magnetic effect (see Lenz's law). *EMF* stands for Electro-Motive Force.

INDUCED VOLTAGE - Voltage created by the action of a magnetic field. Also called *induced EMF*.

Conservation of Energy

INDUCED CURRENT - Current created by the action of magnetic field.

EFFECTIVE RESISTANCE, R_e - Imaginary resistor that substitutes for two or more resistors in a circuit yet leaves the voltage and current unchanged.

VOLTAGE DROP - The voltage (p.d.) between the ends of a resistor in a network of resistors. Also, the work done on one coulomb of charge as the coulomb goes from one end of the resistor to the other.

FORMULAS

a. $PE + KE = $ constant (rising and falling objects, no friction)

b. For resistors in series:

$I_1 = I_2 = I_3 = \cdots = I_n$
$V_s = V_1 + V_2 + V_3 + \cdots + V_n$
$R_e = R_1 + R_2 + R_3 + \cdots + R_n$

c. For resistors in parallel:

$I_{main} = I_1 + I_2 + I_3 + \cdots + I_n$
$V_1 = V_2 = V_3 = \cdots = V_n$
$1/R_e = 1/R_1 + 1/R_2 + \cdots + 1/R_n$

(I_{main} is the current in the *main line* of the circuit, before and after the current splits into branches. V_s is the voltage of the source.)

d. $c_1 m_1 \Delta T_1 = c_2 m_2 \Delta T_2$ e. $V = BLv$

f. $V = \Delta\emptyset/\Delta t$ where $\emptyset = BA$ (see chapter ten, problem four)

g. $E = mc^2$

13.2   THINGS TO KNOW

(1) In the formula $PE = mgh$ (for the gravitational PE of an object near earth - see chapter twelve, Formulas, item *e*) the symbol *h* represents the vertical distance between the object and the base

level, irrespective of the path taken by the object in going to or from the base level (figure 13-1).

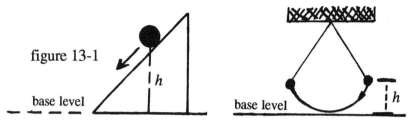

(2) As electrons make their way through a circuit, the electric *PE* of the battery (derived from the battery's chemical energy) is converted to heat, light and other forms. The work done by the battery is equal to the lost electric *PE*. This amount is also equal to the energy provided by the circuit in the form of heat, light, etc.

(3) The formula $V_s = V_1 + V_2 + \cdots + V_n$ and $V_1 = V_2 = \cdots = V_n$ for voltage drops in series and parallel, respectively, follow from the principle of conservation of energy. The formula $I_1 = I_2 = \cdots = I_n$ and $I_{main} = I_1 + I_2 + \cdots + I_n$ for currents in series and parallel, respectively, follow from the principle of conservation of charge.

(4) As resistors are added in parallel with other resistors, the effective resistance of the circuit, R_e, decreases and the main line (total) current increases.

(5) Since heat is a form of energy (internal energy transferred) it obeys the dictates of the law of conservation of energy. When internal energy is transferred from one object to another, and no other forms of energy are involved, the internal energy lost by one object is equal to the internal energy gained by the other. There is, however, no law that temperature is conserved. The number of degrees lost by one object is not necessarily equal to the number of degrees gained by the other. Temperature does not represent an amount of energy but the *average KE* of the molecules of an object.

Conservation of Energy

(6) In the equation $c_1 m_1 \Delta T_1 = c_2 m_2 \Delta T_2$ both ΔT are designed to be positive, the higher temperatue minus the lower temperature. If an object cools down, ΔT is $(T_i - T_f)$; if it warms up, ΔT is $(T_f - T_i)$.

(7) The current in *amperes* and power in *watts* of an induced current can be determined from the induced voltage, V, by making use of Ohm's law, $V = IR$, and the formula $P = VI$ (see chapter twelve).

(8) Lenz's law can be visualized schematically as in figure 13-2.

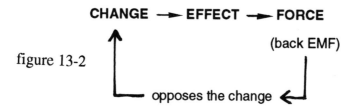

figure 13-2

For example, if a wire segment (in contact with a loop to form a complete circuit) is moved across magnetic field lines (the *change*), as illustrated in figure 13-3, a current is induced in the wire (the *effect*). This current, in turn, leads to a magnetic force (the *back EMF*) that fights the motion of the wire (see chapter ten). This force, if not checked, will stop the motion of the wire and, when that happens, the induced current will die out. To keep the current flowing, an outside agent must counteract the *back EMF* with an

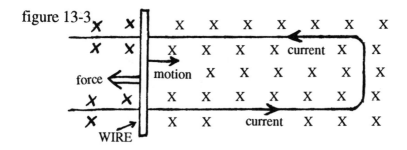

figure 13-3

equally strong but oppositely directed force in order that a state of equilibrium (net force of zero) is maintained. Then the wire continues to move at constant speed (no net force implies no accelerations or decelerations) and the current continues to flow at a constant rate. The energy output provided by the current thus requires an input of work (by an outside agent). The outside agent's loss of energy (doing work implies losing energy) is equal to the energy provided by the current, and energy is conserved.

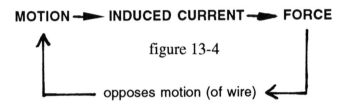

figure 13-4

(9) It is incorrect to say "mass is converted to energy" or "energy is converted to mass". No such conversion occurs, since energy is associated with mass and mass is a form of energy. Instead, think of *mass-energy* as the conserved quantity. The mass-energy of a system remains unchanged.

13.3 SOLVING THE PROBLEMS

(1) A pendulum is set swinging by pulling it through an angle of 30° from the vertical and releasing it. The string is 50 cm long. What is the speed of the bob at the lowest point of the swing? Assume no friction.

Conservation of Energy

SKETCH AND IDENTIFY

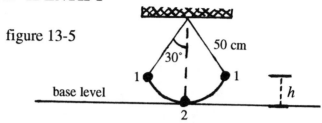

figure 13-5

As the pendulum swings back and forth, its total mechanical energy, $KE + PE$, is conserved. This must be so since other forms of energy (such as heat generated by friction) don't enter into consideration and it is a fundamental law of nature that energy is conserved. Thus, the quantity $mgh + mv^2/2$ remains constant throughout the swing, including the lowest point and the highest point. (True, the earth is part of the *system*, since it exerts a force on the pendulum, and any application of the law of conservation of energy must take its energy into account. But the earth's change in speed is so much smaller than the pendulum's that any gain or loss in the earth's KE is insignificant compared to that of the pendulum and can safely be ignored.)

If the highest point on either side of the swing is denoted by the subscript *one* and the lowest point by *two*, we may claim that:

$$mgh_1 + mv_1^2/2 = mgh_2 + mv_2^2/2$$

Since the bob of the pendulum is at rest whenever it gets to either of the highest points - it changes direction there - the quantity $mv_1^2/2$ is zero. If the base level is chosen to be the horizontal line running through the lowest point of the swing, as illustrated in figure 13-5, the quantity mgh_2 is zero since h is zero at the base level. The equation above can therefore be simplified as follows:

$$mgh_1 = mv_2^2/2$$

The mass of the swinging bob need not be known, since the m's drop out of both sides of the equation. (This also implies that the connection between the speed at the lowest point and the height at the highest point - and the solution to our problem - does not depend on the mass.) Little g is known and h_1 can be determined from the geometry of the problem. The only remaining unknown is therefore v_2, and it can be obtained by solving the above equation.

LINK UNKNOWN TO GIVEN

First we determine h_1 from the diagram in figure 13-6.

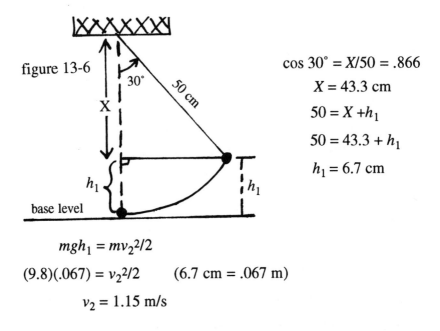

figure 13-6

$\cos 30° = X/50 = .866$

$X = 43.3$ cm

$50 = X + h_1$

$50 = 43.3 + h_1$

$h_1 = 6.7$ cm

$$mgh_1 = mv_2^2/2$$
$$(9.8)(.067) = v_2^2/2 \quad (6.7 \text{ cm} = .067 \text{ m})$$
$$v_2 = 1.15 \text{ m/s}$$

(2) The pulley system illustrated below (figure 13-7) is used to lift a 1,000 N weight. To lift the weight at a steady pace, a 400 N strong pull must be exerted at the end of the rope. How much heat energy is generated by friction between the ropes and wheels as the weight is so lifted through a distance of two meters?

Conservation of Energy

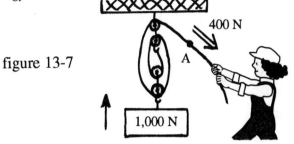

figure 13-7

SKETCH AND IDENTIFY

A key point about pulley systems is that the distance through which the exerted force moves is not the same as the distance through which the weight is lifted. If point A on the rope (figure 13-7), for example, is pulled through a distance of one meter, then one meter of rope has been removed from between the wheels. This means that each of the four rope segments situated between the upper and lower sets of wheels has been shortened by one fourth of a meter and that, consequently, the weight was lifted - not one meter, but - one fourth of a meter.

This imbalance in distance forces nature into a corner. In order to conserve energy it is necessary that the work done pulling the rope (energy lost by whoever or whatever is doing the pulling) be equal to the gain in PE achieved by lifting the weight - assuming no friction and therefore no heat generated, and a constant pace that leaves the KE of the weight, wheels and rope unchanged. Now, this can only be the case if the force exerted in pulling the rope is, in our example, one fourth as strong as the weight, that is, 250 N. As the object rises one fourth of a meter its gain in PE is equal to:

$$mgh = Wh = (1000)(.25) = 250 \text{ J}$$

and the work done (and energy lost) by the person or machine pulling the rope is:

$$W = Fd\cos\theta = (250)(1)(1) = 250 \text{ J}$$

and energy is conserved. This is the reason why pulley systems are widely used in construction projects - 1,000 N weights can be lifted with only 250 N of force!

In our problem, however, friction does exist. This necessitates the exertion of more than 250 N of force to lift the weight at a steady pace. The extra 150 N of force specified in our problem (for a total of 400 N) is needed to overcome friction which acts to oppose the rubbing of ropes against wheels, wheels against axles, and so on. Nevertheless, energy is still conserved, as it always must be. But with friction in the picture another form of energy enters into our calculations - heat generated by rubbing.

LINK UNKNOWN TO GIVEN

Energy lost = Energy gained

Work done = PE + Heat

$Fd\cos\theta = mgh$ + Heat   (object starts at base level)

$(400)(8)(1) = (1000)(2)$ + Heat   (force moves 8 m as weight rises 2 m)

Heat = 1200 J

(3) In the circuit diagrammed below (figure 13-8), three resistors are connected in series to a 24-volt source. Resistor *one* has 4 ohms of resistance, resistor *two* is a rheostat - its resistance can be adjusted by turning a knob, and resistor *three* has 6 ohms of resistance. (a) What should be the resistance of the rheostat in order that the current through the circuit be 2 amperes? (b) Resistor *two* is immersed in 100 grams of water and the heat generated in it by the current is completely transferred to the water. If the 2-amp current flows for three minutes and the initial temperature of the water is 30°C, what is the final temperature of the water?

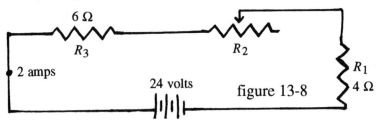

figure 13-8

SKETCH AND IDENTIFY

We label the resistance of the rheostat that leads to a current of 2 amperes, R_2. The three resistors in series can be replaced by one *effective resistor*, R_e, without changing the current, if R_e is chosen carefully - it must obey the formula $R_e = R_1 + R_2 + R_3$. An imaginary circuit with a single resistor so chosen, connected to the same 24-volt source, will allow the same current of 2 amperes to flow through it. Since such a circuit would obey Ohm's law, $V = IR$, we are in a position to compute the resistance of that single resistor, R_e. Then, from the equation $R_e = R_1 + R_2 + R_3$ the resistance of the rheostat, R_2, can be computed.

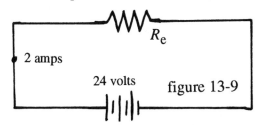

figure 13-9

The next step is to compute the *voltage drop* for resistor *two*, V_2, by applying Ohm's law to resistor *two* ($V_2 = I_2 R_2$). The power of resistor *two* can then be determined by applying the formula $P = VI$ to resistor *two* ($P_2 = V_2 I_2$). This power is the rate at which electrical PE is converted to heat in resistor *two*. The heat energy generated by resistor *two* in three minutes can then be found from $E = I^2 Rt$ or $E = Pt$. Since this heat is transferred entirely to the water, it is the value of Q in the formula $Q = cm\Delta T$ when that formula is applied to the water. Use of that formula, in turn, leads to knowledge of ΔT and the water's final temperature, since we know c, the specific heat of water (4.18 J/gm·°C) and m, the mass of the water (100 grams).

LINK UNKNOWN TO GIVEN

(a) $V = IR_e$ (applied to imaginary circuit, figure 13-9)

$24 = (2)R_e$

$R_e = 12 \, \Omega$

$R_e = R_1 + R_2 + R_3$

$12 = 4 + R_2 + 6$

$R_2 = 2 \, \Omega$

(b) $V_2 = I_2 R_2$

$V_2 = (2)(2) = 4$ Volts

(From $V_1 = I_1 R_1 = (2)(4)$ we determine that V_1 is 8 volts and from $V_3 = I_3 R_3 = (2)(6)$ we know that V_3 is 12 volts. Do the three voltage drops add up to the voltage of the source, V_s, as dictated by the formula $V_s = V_1 + V_2 + V_3$? Yes, since $8 + 4 + 12$ is equal to 24.)

$P_2 = V_2 I_2$

$P_2 = (4)(2) = 8$ Watts $= 8$ J/s

Resistor *two* generates 8 Joules of heat energy per second. After three minutes (180 seconds) of operation the total amount of heat energy provided by resistor *two* is:

$E = Pt$	OR	$E = I^2 R t$
$E = (8)(180)$		$E = (2)^2(2)(180)$
$E = 1440$ J		$E = 1440$ J

Conservation of Energy

$Q = cm\Delta T$

$1440 = (4.18)(100)(\Delta T)$ (units: joules, grams, °C)

$\Delta T = 3.44$ °C

The final temperature of the water, T_f, is determined as follows:

$T_f = T_i + \Delta T = 30 + 3.44 = 33.44$ °C

(4) In the series-parallel combination circuit illustrated below (figure 13-10), find the voltage drop across, current through and power of each of the five resistors.

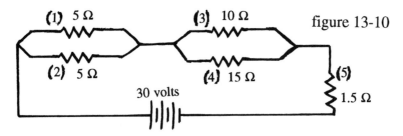

figure 13-10

SKETCH AND IDENTIFY

The key to solving network problems such as this is to reduce the number of resistors to one. This is done by replacing many resistors with one so called *effective resistor* whose resistance is labeled R_e. Resistors in parallel are replaced by an R_e as prescribed by the formula $1/R_e = 1/R_1 + 1/R_2 + \cdots + 1/R_n$. Resistors in series are replaced by an R_e via the equation $R_e = R_1 + R_2 + \cdots + R_n$.

These formulas for replacement resistors are designed to accomplish specific purposes and it is important to keep these in mind. When resistors in parallel are replaced by one effective resistor via the formula $1/R_e = 1/R_1 + 1/R_2 + \cdots + 1/R_n$, the voltage drop across the single effective resistor is the same as the voltage drop across each of the parallel resistors it replaced - no matter what else is connected to the circuit. When resistors in series are replaced by one effective resistor whose resistance is provided by the formula $R_e = R_1 + R_2 + \cdots + R_n$, the current through the single effective resistor is the same as the current through each of the resistors in series it replaced.

Sometimes a sequence of replacements is necessary. The five resistors in the problem confronted here cannot be replaced, in one bold stroke, with one effective resistor because they are not all either in series or in parallel with each other. The effective resistor formulas specified above are applicable only to groups of resistors whose members are all in series or all in parallel. So we proceed in steps, working our way from the inside out. First resistors *one* and *two*, in parallel, are replaced by an effective resistor which we designate R_{e1}. Then resistors *three* and *four*, also in parallel, are replaced by another effective resistor which we designate R_{e2}. Then we replace the two R_e's so obtained and resistor *five*, all of which are in series, with yet another effective resistor, called R_{e3}. This final R_e is imagined to be connected to our 30-volt source, forming a "one resistor, one source" circuit.

A schematic of this sequence of replacements by effective resistors is sketched in figure 13-11.

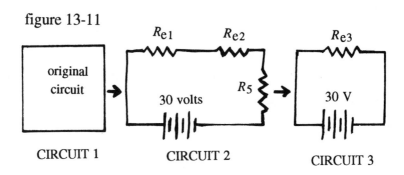

figure 13-11

CIRCUIT 1   CIRCUIT 2   CIRCUIT 3

Next, Ohm's law is applied to the single-resistor circuit to obtain the current through R_{e3}. This current is equal to the current through each of the three resistors in series it replaced - R_{e1}, R_{e2} and R_5. Then Ohm's law is applied again, this time to each resistor in circuit *two* (figure 13-11), to determine the voltage drop across R_{e1}, R_{e2} and R_5. Each of these voltage drops is, in turn, identical to the voltage drop across each of the branch resistors in parallel in the original circuit it replaced.

Armed with knowledge of the voltage drop across every resistor in our original circuit, we next determine each resistor's current and power by applying Ohm's law and the formula $P = VI$ to each.

LINK UNKNOWN TO GIVEN

Resistors *one* and *two* in parallel are replaced by an effective resistor, R_{e1}, as follows:

$1/R_e = 1/5 + 1/5$

$1/R_e = 2/5$

$2R_e = 5$

$R_{e1} = 2.5 \ \Omega$

Similarly, resistors *three* and *four* in parallel are replaced by another effective resistor, R_{e2}, as such:

$1/R_e = 1/10 + 1/15$

$1/R_e = 3/30 + 2/30 = 5/30$

$5R_e = 30$

$R_{e2} = 6 \ \Omega$

We now have the following series circuit (figure 13-12):

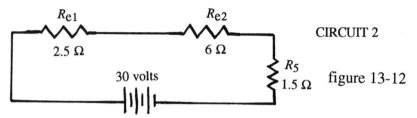

CIRCUIT 2

figure 13-12

These three resistors in series are now replaced by yet another effective resistor, R_{e3}, which reduces the number of resistors to one.

$R_{e3} = R_{e1} + R_{e2} + R_5$

$R_{e3} = 2.5 + 6 + 1.5 = 10 \, \Omega$

The final circuit looks like this (figure 13-13):

CIRCUIT 3

figure 13-13

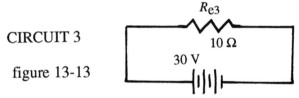

Ohm's law reveals the current through this single-resistor circuit.

$V = IR$

$30 = I(10)$

$I = 3 \, A$

Now we begin to work our way backward to the original circuit. If three amperes is the current through the single-resistor circuit, then three amperes must be the current through each of the three resistors in series replaced by the single resistor, R_{e3} (figure 13-14):

figure 13-14

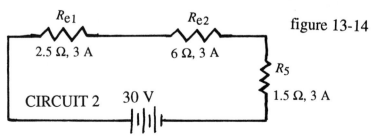

Conservation of Energy

The voltage drop across each of these resistors is now determined via Ohm's law, $V = IR$.

$V_{e1} = (3)(2.5) = 7.5$ Volts, $V_{e2} = (3)(6) = 18$ Volts, $V_5 = (3)(1.5) = 4.5$ Volts.

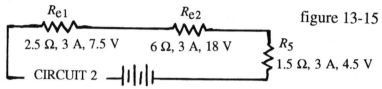

figure 13-15

(The formula $V_s = V_1 + V_2 + V_3$ is obeyed by this circuit, as it must be by any series circuit, since 30 equals $7.5 + 18 + 4.5$)

The voltage drop across each of the two R_e's in circuit *two* must be the same as the voltage drop across each of the branch resistors in parallel it replaced. Returning to our original circuit and putting in all the data gathered so far, we obtain the following (figure 13-16):

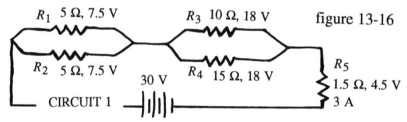

figure 13-16

The current through R_1, R_2, R_3 and R_4 can now be determined from Ohm's law:

$V_1 = I_1 R_1$	$V_2 = I_2 R_2$	$V_3 = I_3 R_3$	$V_4 = I_4 R_4$
$7.5 = I_1(5)$	$7.5 = I_2(5)$	$18 = I_3(10)$	$18 = I_4(15)$
$I_1 = 1.5$ A	$I_2 = 1.5$ A	$I_3 = 1.8$ A	$I_4 = 1.2$ A

The sum of the currents through R_1 and R_2, $1.5 + 1.5$, is equal to the main-line current of three amperes, as is the sum of the currents through R_3 and R_4, $1.8 + 1.2$ amperes. This is always true for branch resistors in parallel.

The power through each resistor can now be obtained by applying $P = VI$ to each:

$P_1 = V_1 I_1 = (7.5)(1.5) = 11.25$ W

$P_2 = V_2 I_2 = (7.5)(1.5) = 11.25$ W

$P_3 = V_3 I_3 = (18)(1.8) = 32.4$ W

$P_4 = V_4 I_4 = (18)(1.2) = 21.6$ W

$P_5 = V_5 I_5 = (4.5)(3) = 13.5$ W

Putting all our results together, we arrive at the following schematic (figure 13-17) which contains all the information there is to know regarding each of the five resistors in our original circuit. Notice that all the pieces of the puzzle fit together, as they must. The currents in parallel branches add up to the main current, the voltage drops in series add up to the source voltage, and every resistor obeys Ohm's law.

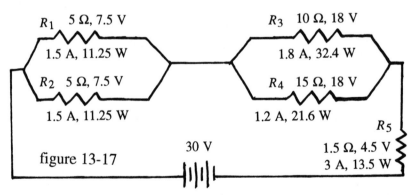

figure 13-17

(5) A wire segment is situated between two opposite magnetic poles, as illustrated in figure 13-18. The wire is made to move into the page (away from the reader) at the constant rate of 2 m/s. The magnetic field intensity between the poles is 8 tesla and the length of the moving wire segment in the field is 10 cm. The wire segment is in contact with a conducting loop to form a complete circuit. The resistance of the circuit is 0.5 ohms. (a) What is the direction of the

Conservation of Energy

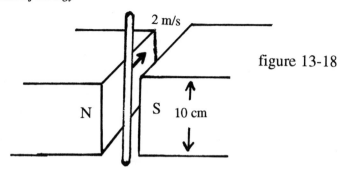

figure 13-18

induced current in the wire segment? (b) What is the magnitude of the induced current in the circuit? (c) How strong a force must be exerted to maintain the speed of the wire? (d) How much work must be done per second to maintain the speed of the wire? (e) How much energy is provided by the circuit per second? (f) Compare your answers to parts *d* and *e*. What is the significance of this comparison?

SKETCH AND IDENTIFY

As the wire moves across the magnetic field lines - which are directed perpendicularly to the poles, from the N-pole to the S-pole - the electrons in the wire are moving along with it. Those electrons are therefore moving perpendicularly to magnetic field lines and, according to the rules of magnetism, a force is exerted by the field on the electrons. The direction of this force is provided by hand rule three (chapter ten, Formulas, item *g*). This force acts to push the electrons parallel to and through the wire and leads to an induced current. The direction of the induced current in the wire is therefore the same as the direction of the magnetic force exerted on the electrons. (The protons in the wire also move along with the wire, perpendicularly to the field lines, and consequently the magnetic field exerts a force also on them. But this force does not lead to a current, since protons are not free to move through the wire.)

This induced current is associated with an induced voltage, V, whose magnitude is provided by the formula $V = BLv$. (Care must be taken in using this formula. The upper case V represents the induced voltage, while the lower case v represents the velocity of the wire.) Ohm's law, $V = IR$, yields the magnitude of the induced current, I, once we know V and R.

The existence of the induced current gives birth to yet another magnetic force, since any magnetic field exerts a force on a current that flows perpendicularly to its field lines. This force is directed opposite to the motion of the wire - out of the page and toward the reader. This can be ascertained either by again applying hand rule three or by making use of Lenz's law (Things to Know, item eight). To keep the wire moving at constant speed (and the current flowing at a steady rate) it is necessary that the wire be in a state of equilibrium - the net force on it must be zero. Only then will the wire neither accelerate nor decelerate. This implies that an outside agent (you, me, a steam engine, whatever) must exert an equally strong force in the direction of the wire's motion - into the page.

The magnitude of the force exerted by the field on the current out of the page is provided by $F = BIL$ (chapter ten, Formulas, item a). This is therefore how strong the outside agent's counter-acting force into the page must be to maintain the wire's speed.

The work done by the outside agent per second can be determined from $W = Fd\cos\theta$ where d is the distance traveled by the wire per second. Since the outside agent must act in the direction of the wire's motion, F and d are in the same direction and $\cos\theta = 1$. The distance traveled by the wire per second is given - it is the speed of the wire.

The energy provided by the circuit per second is the power of

Conservation of Energy 287

the circuit, in watts. It is provided by $P = VI$ and we know both V (the induced voltage) and I (the induced current). If energy is conserved, the energy provided by the circuit per second should equal the work done by the outside agent per second, which is the energy lost by the outside agent per second. Let us see if this is indeed the case.

LINK UNKNOWN TO GIVEN

(a) Using hand rule three, we proceed as follows, with the left hand:

    Index finger - field intensity: $\longrightarrow$
    Thumb - motion of wire and electrons: X X X   fig. 13-19
    Middle finger - force on electrons: $\uparrow$

The induced current is directed upward (toward the top of the page) in the wire segment that moves across the field lines.

(b) $V = BLv$

$V = (8)(.1)(2) = 1.6$ Volts   (10 cm = .1 m)

$V = IR$

$1.6 = I(.5)$

$I = 3.2$ Amperes

(c) $F = BIL$

$F = (8)(3.2)(.1) = 2.56$ N

To verify that the force exerted by the field on the current is directed out of the page, toward the reader, we use hand rule three, once again with the left hand, as follows:

    Index finger - field intensity: $\longrightarrow$
    Thumb - current of electrons: $\uparrow$ figure 13-20
    Middle finger - force: • • •

A force of 2.56 N must be exerted on the wire by an outside

agent, in the direction of the wire's motion, in order to counter-act the magnetic force of 2.56 N on the wire in the opposite direction. The net force on the wire then is *zero*, and the wire continues to move into the page, at constant speed. This keeps the induced voltage (which depends on the speed) and the induced current (which depends on the induced voltage) stable and constant.

(d) $W = Fd\cos\theta$

$W = (2.56)(2)(1) = 5.12$ J   (per second)

(e) $P = VI$

$P = (1.6)(3.2) = 5.12$ watts $= 5.12$ J/s

(f) The work done by the outside agent per second in maintaining the motion of, and the current through, the wire is identical to the energy provided by the circuit per second. Energy is therefore conserved - what is produced by the current (in heat, light and other forms) is lost by the outside agent, every second.

(6) A 200 gram piece of copper at 30 °C is immersed in 100 grams of hot water at 100 °C. Assuming no heat is lost to the surroundings, what is the final temperature of the mixture? (The specific heat of water is 4.18 J/gm·°C and that of copper is .39 J/gm·°C.)

SKETCH AND IDENTIFY

When hot meets cold, internal energy is transferred from the hot to the cold. As a result, the hot object's temperature decreases and the cold object's temperature rises. This process continues until the two temperatures become equal. Then neither body is warmer than the other and the flow of energy comes to an end.

The law of conservation of energy dictates that the internal

Conservation of Energy

energy gained by one object (the originally colder one) be equal to the internal energy lost by the other object (the originally hotter one). The internal energy change (gain or loss) experienced by an object, symbolized by Q, is related to the change in temperature of the object, ΔT, by the formula $Q = cm\Delta T$. Since the internal energy gain of one object is equal to the internal energy loss of the other we may write:

$$c_1 m_1 \Delta T_1 = c_2 m_2 \Delta T_2$$

The ΔT of an object is the difference between its initial and final temperatures. The final temperature in our problem is the same for both objects, and will be symbolized by T_f (temp-final). To render ΔT positive on both sides of the equation, the lower temperature is subtracted from the higher temperature. For the originally hotter object, whose temperature started high and *went down* to the final temperature, ΔT is $T_h - T_f$ (temp-hot minus temp-final). For the originally colder object, whose temperature started low and *went up* to the final temperature, ΔT is $T_f - T_c$ (temp-final minus temp-cold). We thus write:

$$c_h m_h (T_h - T_f) = c_c m_c (T_f - T_c)$$

where the subscript h represents the originally hotter object (the water) and the subscript c represents the originally colder object (the copper).

LINK UNKNOWN TO GIVEN

$$c_h m_h (T_h - T_f) = c_c m_c (T_f - T_c)$$
$$(4.18)(100)(100 - T_f) = (.39)(200)(T_f - 30) \quad \text{(joules, gms, °C)}$$
$$(418)(100 - T_f) = (78)(T_f - 30)$$
$$(5.36)(100 - T_f) = T_f - 30$$

$$536 - (5.36)T_f = T_f - 30$$

$$566 = (6.36)T_f$$

$$T_f = 89\ °C$$

Notice that the changes in temperature are not equal. The copper's increase in temperature is (89 - 30) or 59 degrees, while the water's drop in temperature is much smaller, (100 - 89) or 11 degrees (figure 13-21).

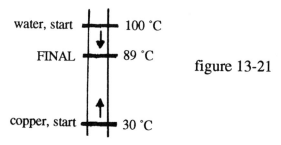

figure 13-21

(7) The circularly shaped wire in the plane of figure 13-22 is situated in a magnetic field whose intensity is directed into the page. The resistance of the wire is .04 Ω, and its diameter is 20 cm. If the magnitude of the field's intensity increases at the rate of 2 tesla per second, (a) what is the direction of the current induced in the wire loop, (b) what is the induced voltage, and (c) what is the magnitude of the current induced in the wire?

figure 13-22

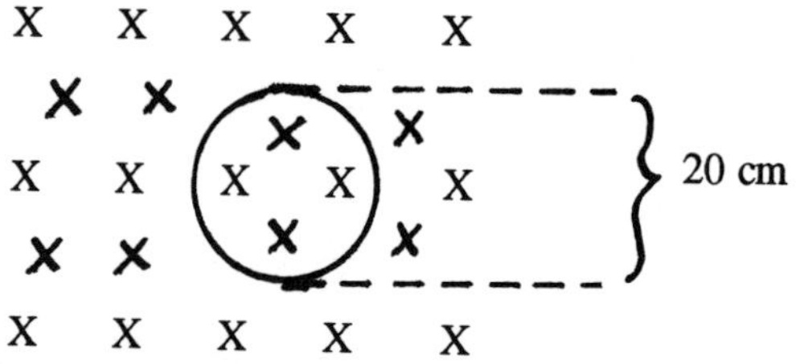

SKETCH AND IDENTIFY

Current and voltage are induced in this circular wire, not because it moves across magnetic field lines, but because the intensity of the magnetic field enclosed by the loop is changing over time. A

Conservation of Energy

changing magnetic field creates an electric field, and electric fields (unlike magnetic fields) exert forces on charges - such as the electrons in the wire - even if those charges are not moving. This force acts to propel the electrons through the wire, thereby inducing a current and a voltage.

The magnitude of the induced voltage, V, can be calculated from Faraday's law, $V = \Delta\emptyset/\Delta t$, where $\emptyset$ is the number of field lines enclosed by the loop. That number is equal to the density of field lines per unit area, provided by the value of B (webers per square meter), times the area bounded by the loop. In other words, $V = \Delta(BA)/\Delta t$. Since A, the area bounded by the loop, is constant in our problem, we may write: $V = (A)\Delta B/\Delta t$.

The magnitude of the induced current can be obtained from the induced voltage by making use of Ohm's law, $V = IR$. The resistance R, the area A and the ratio $\Delta B/\Delta t$ can all be obtained from the data provided in the problem.

To find the direction of the induced current we must resort to Lenz's law:

figure 13-23

The *change* in this problem is the increasing intensity of the magnetic field directed into the page. The *effect* is the induced current. This induced current gives birth to its own magnetic field (chapter ten) which, according to Lenz's law, opposes the *change*. This can only mean that the field created by the current is directed out of the page, thereby acting to diminish the increase in the field's intensity into the page.

The direction of a circular current is related to the direction of the magnetic field it creates via hand rule two (chapter ten, Formulas, item f). Since the field created by the induced current in this problem must be directed out of the page, we are in a position to determine the direction of the current by working our way backward through that hand rule.

LINK UNKNOWN TO GIVEN
(a) Applying hand rule two:
Four fingers - current of electrons: figure 13-24
Thumb - field inside loop: • • • •

Since the induced current in our problem must create a magnetic field whose intensity is directed out of the page, that current must flow clockwise. (We concern ourselves only with the field inside the wire loop. The changing field outside the loop does not lead to an induced current, as is evident from the formula $V = \Delta\emptyset/\Delta t$ and the definition of $\emptyset$.)

(b) $V = (A)\Delta B/\Delta t$
$V = (\pi)(.20/2)^2(2)$ (20 cm = .20 m)
$V = .063$ Volts

(c) $V = IR$
$.063 = I(.04)$
$I = 1.57$ Amperes

13.4   ON YOUR OWN PROBLEMS

(Asterisk indicates solution appears in appendices.)

*(1) An object is launched vertically upward with a velocity of 100 m/s. What is its velocity when it reaches a height of 200 meters above the launching point?

Conservation of Energy

*(2) Fifty grams of a metal at 200 °C are immersed in 100 grams of water at 30 °C. The final temperature of the mixture is 40 °C. Assuming no heat loss to the environment, what is the specific heat of the metal?

*(3) A 100 kg object slides down an incline. It starts at rest at a height of 5 meters and arrives at the bottom of the incline with a velocity of 3 m/s. How much heat energy was generated by the slide, if any?

*(4) Determine the voltage drop across, current through and power of each of the four resistors in the network illustrated in figure 13-25.

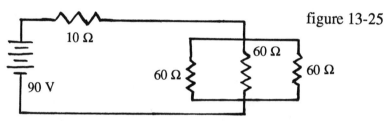

figure 13-25

*(5) A straight wire segment slides over a long U-shaped wire situated in a magnetic field, as illustrated (figure 13-26). Good metal-to-metal contact is maintained at both points where the wires meet, such that electrons can easily travel from wire to wire. The magnitude of the field's intensity is 8 tesla, the resistance of the circuit is 0.5 Ω, the wire segment moves at a constant rate of 4 m/s,

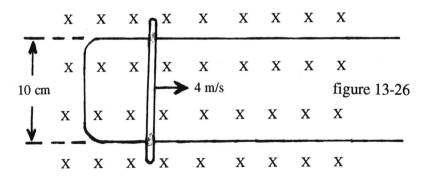

figure 13-26

and the length of the moving wire segment in the field is 10 cm. (a) What is the direction and magnitude of the current induced in the circuit? (b) How strong a force must be exerted on the wire to keep it moving at constant speed? (c) How much work does this force do in one minute? (d) Show that the energy lost by the agent exerting this force is just equal to the energy generated by the circuit during the same one minute period.

(6) A 400 kg cart is rolling along a frictionless roller-coaster ride, as illustrated in figure 13-27. When the cart is at point A, which is 50 meters above ground level, its speed is 12 m/s. (a) What is the speed of the cart at point B, which is 20 meters from the ground? (b) Does the cart make it, on its own, up the hill to point C, which is 80 meters from the ground?

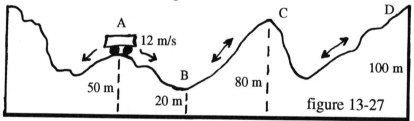

figure 13-27

(7) The cart in problem six is now given a boost in order that it may arrive at point D, which is 100 meters above ground level. How much work must the motor providing the boost do in order to accomplish this?

(8) A railroad car with mass 3000 kg rolls along a level track at a speed of 20 m/s. It collides with a stationary car of equal mass, whereupon the cars become locked together. (a) What is the speed of the two-car combination immediately after the collision? (b) How much *KE* is lost as a result of the collision? (c) What might have happened to this energy?

(9) The lever in figure 13-28 is used to lift a 1000 N rock. For

every cm the rock rises, at one end of the lever, the other end of the lever, where the force is applied, moves 8 cm. How strong must the force be to lift the rock at a steady pace?

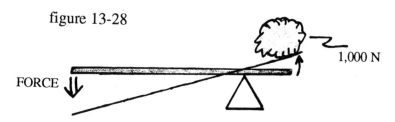

figure 13-28

FORCE

1,000 N

(10) Three 15 Ω resistors are connected in parallel. This arrangement is connected in series with a 10 Ω resistor. The entire circuit is then connected to a 90-volt difference in potential. (a) What current flows thru each resistor? (b) What is the power of each resistor?

(11) A vertical wire segment in contact with a conducting loop is moving from east to west in a magnetic field whose intensity is directed from south to north. The wire is moving at the constant rate of 6 m/s, the magnetic field intensity is 30 W/m^2, the length of the moving wire in the field is 20 cm and the resistance of the circuit is 0.1 Ω. (a) What is the direction of the induced current in the wire segment? (b) How much current flows thru the circuit? (c) How strong a force must be exerted to maintain the speed of the wire? (d) How much work must be done per second to maintain the speed of the wire? (e) How much energy is generated by the circuit per second? (f) Compare your answers to parts d and e. What is the significance of this comparison?

(12) A cardboard tube is wound with two windings of insulated wire. The inner winding, A, is connected to a battery, a switch and a rheostat, while the outer winding, B, is connected only to a fixed resistor, R. The length of the tube and the fixed resistor are both

oriented horizontally, in the right-left direction, parallel to each other. In each of the following situations state whether the induced current in winding B, if any exists, is directed from left to right or right to left. (a) The current in winding A is from left to right and increasing. (b) The current in winding A is from right to left and decreasing. (c) The current in winding A is from left to right and decreasing. (d) The current in winding A is from right to left and unchanging.

(13) A 2 kg block of lead at 90 °C comes in contact with a 3 kg block of aluminum at 20 °C. Both are placed in an evacuated vessel so that no heat is lost to the environment. What is their final temperature? (The specific heat of lead is 0.031 calories/gm·°C and that of aluminum is 0.217 cal/gm·°C.)

(14) A 4 kg block is dropped from a height of 6 meters onto a spring whose constant is 200 N/m. What is the maximum compression of the spring?

(15) A 5 kg projectile is launched from ground level with a velocity of 100 m/s. Calculate the total energy of the projectile at the highest point in its trajectory for each of the following launching angles: 30°, 45°, 60° and 90°. What is the significance of this comparison? (Do not base calculations on energy conservation.)

(16) Three equal resistors are connected in series. When a certain potential difference is applied across this combination, the total power generated is 100 watts. What would the total power be if the same resistors were connected in parallel with the same p.d. across the parallel combination?

(17) Two lamps marked "100 watts, 120 volts" and "60 watts, 120 volts" are connected in series across a 120-volt potential difference. What is the power generated by each lamp, assuming the

resistance of each does not significantly depend on the temperature and current?

(18) A 50 gram sample of an unknown material at a temperature of 125 °C is dropped into a calorimeter which consists of 100 grams of water resting in a copper can of mass 200 grams. The initial temperature of the calorimeter is 20 °C and the final temperature of all three materials is 25 °C. What is the specific heat of the unknown material? Assume no heat loss to the environment. (The specific heat of copper is 0.39 J/gm·°C and that of water is 4.18 J/gm·°C.)

(19) A bar magnet is dropped through a wire loop as shown in figure 13-29. As the north pole approaches the loop, current is induced in the loop. What is the direction of this electron flow from the point of view of someone looking down at the loop?

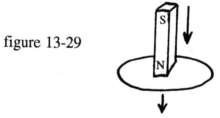

figure 13-29

(20) In the nuclear fusion reaction $_1H^2 + {_1H^3} \rightarrow {_2He^4} + {_0n^1}$ the combined mass of the reactants is 5.03013 amu and the combined mass of the products is 5.01127 amu. (One *amu* - atomic mass unit - is equal to 1.66 x 10^{-27} kg.) How much energy is released as a result of each such reaction?

(21) A woman uses a wheel-and-axle system to lift a 500 N weight to a height of 6 meters, as illustrated in figure 13-30. The radius of the wheel is 20 cm and the radius of the axle is 5 cm. The wheel is rigidly fastened to the axle, so that both must turn together. (a) Assuming frictional effects may be ignored, how strong a force must the woman exert on the rope (wound around the wheel) in

order that the weight rises at a steady pace? (b) Through how many meters must she pull the rope? (c) How much work does she do? (d) How much energy does she lose? (e) Where and in what form is the energy after she lost it? (f) Show that the energy lost by the woman is equal to the energy gained in its new form?

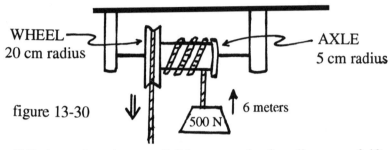

figure 13-30

(22) An archer draws a 0.2 kg arrow back a distance of 40 cm. To prevent the arrow from flying out he must exert a force of 80 N. (a) Assuming the bow obeys Hooke's law ($F = kx$ - chapter twelve, problem six), how much potential energy is stored in the bow-and-arrow in this condition? (b) What is the speed of the arrow immediately after it is released? (c) If the arrow is fired vertically upward, to what height does it rise?

(23) Give an explanation based on physics principles for the following statements: (a) The handles of metal-cutting shears are made longer than the cutting blades, but the handles of tailors' scissors are made shorter than the cutting blades. (b) No device or machine saves work.

14

CHAPTER FOURTEEN

WAVES AND PARTICLES

14.1 VOCABULARY AND FORMULAS

WORD	SYMBOL	UNIT
Wavelength	λ	meter/cycle
Amplitude	A	see Definitions
Period	T	second
Frequency	f	cycles/sec (hertz, Hz)
Planck's constant	h	joule·sec (J·s)
Stopping potential	V_0	joule
Electron-volt	eV	1 eV = 1.6 x 10^{-19} J
Work function	W	joule
Threshold frequency	f_0	cycles/sec (Hz)

DEFINITIONS

WAVELENGTH - The distance between corresponding points of adjacent cycles (figure 14-1).

figure 14-1
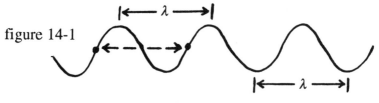

AMPLITUDE - The maximum disturbance in each cycle. For crests and troughs in a rope or body of water, the amplitude is the maximum height (or depth) the rope or water rises (or falls) to in each complete cycle. For sound waves, the amplitude is the maximum deviation in pressure or density from the norm of the surrounding, undisturbed, medium. For electromagnetic (*EM*) waves, including light, the amplitude is the maximum intensity of the electric or magnetic field per cycle. (See also Things to Know, item *six*.)

FREQUENCY (observed) - The number of cycles passing by an observer per second.

FREQUENCY (emitted) - The number of complete cycles generated by the source per second.

PERIOD (observed) - The time it takes one complete cycle to pass by an observer.

PERIOD (emitted) - The time it takes one complete cycle to be generated by the source.

INTERFERENCE - Two or more waves superimposed upon each other at one or more points within the medium through which they travel. The net disturbance at each such point is generally the vector sum of the disturbances generated at the point.

CONSTRUCTIVE INTERFERENCE (*CI*) - Two or more waves superimposed upon each other at a point in such a manner that the magnitude of the resultant disturbance is greater than the magnitude of any of the individual disturbances at that point. For example, a two cm tall crest of one wave (disturbance: +2) overlaps a one cm tall crest of another wave (disturbance: +1). The net result is a three cm tall crest at the point where the crests meet. Since this disturbance (+3) is greater than either of the individual disturbances that overlap at that point, the waves are said to *interfere constructively* with each other at that point.

DESTRUCTIVE INTERFERENCE (*DI*) - Two or more waves superimposed upon each other at a point in such a manner that their disturbances negate each other. For example, a two cm tall crest of one wave (disturbance: +2) overlaps a two cm deep trough (disturbance: -2) of another wave. The net result is a disturbance of *zero* at the point where these disturbances meet. These waves are said to *interfere destructively* with each other.

WORK FUNCTION - The minimum amount of energy an electron must have in order to be able to escape from a particular material. This is also the minimum amount of energy an escaping electron will lose as it makes its way out of the material.

STOPPING POTENTIAL - The minimum voltage required to turn back all the electrons - even the fastest moving ones - that are ejected from a positively charged photomaterial before they arrive at the opposite, negatively charged, plate.

THRESHOLD FREQUENCY - The lowest frequency of *EM* radiation that can knock electrons out of a particular photomaterial. Different photomaterials have different threshold frequencies and different work functions.

FORMULAS

a. $\lambda f = v$ b. $f = 1/T$ c. $E = hf$ d. $W = hf_0$ e. $p = h/\lambda$
f. $x/L = \lambda/d$ (double slit)   g. $y/L = \lambda/w$ (single slit)
h. $d\sin\theta = m\lambda$ (double slit, bright bands, $m = 0, 1, 2, 3, ...$)
i. $d\sin\theta = (m + 1/2)\lambda$ (double slit, dark bands, $m = 0, 1, 2, 3, ...$)
j. $W\sin\theta = m\lambda$ (single slit, dark bands, $m = 1, 2, 3, ...$)
k. $KE_{max} = hf - W$ ($h = 6.63 \times 10^{-34}$ J·s)   l. $KE_{max} = V_0 e$

14.2 THINGS TO KNOW

(1) Frequency is not synonomous with speed. A faster moving wave does not imply a greater frequency nor does a smaller frequency indicate that the wave is slower moving. Speed is meters traveled per second; frequency is cycles passing by per second. They are not the same.

(2) There are two types of *frequency*, emitted and observed, as defined above. The two frequencies are equal to each other when neither the source nor the observer is moving. When motion does exists on the part of either the source or the observer or both, the two frequencies are not equal to each other (see *doppler effect* below).The formula $\lambda f = v$ is applicable to the observed characteristics (λ, f and v) of a wave.

(3) There are two types of *period*, emitted and observed, as defined previously. The reciprocal relationship that exists between frequency and period ($f = 1/T$) is valid for corresponding types only. That is to say, the emitted frequency is the reciprocal of the emitted period and the observed frequency is the reciprocal of the observed period.

(4) All velocities associated with the doppler effect (chart below) refer to motion from the point of view of the medium through which the wave is propagated, except for the symbol v_o which represents the velocity of the wave from the point of view of the observer. Similarly, the phrase "at rest", "moving toward" and "moving away" all refer to what the medium sees. In addition, the formulae in the chart below are applicable only to situations where the velocity of the source (v_s) and the velocity of the observer (v_r) are smaller than the velocity of the wave relative to the medium (v).

THE DOPPLER EFFECT

MOTION	FREQUENCY	WAVELENGTH	SPEED
(a) Source moves toward observer at rest	$f_o = \left(\dfrac{v}{v - v_s}\right) f_e$	$\lambda_o = \left(\dfrac{v - v_s}{v}\right) \lambda$	$v_o = v$
(b) Source moves away from observer at rest	$f_o = \left(\dfrac{v}{v + v_s}\right) f_e$	$\lambda_o = \left(\dfrac{v + v_s}{v}\right) \lambda$	$v_o = v$
(c) Observer moves toward source at rest	$f_o = \left(\dfrac{v + v_r}{v}\right) f_e$	$\lambda_o = \lambda$	$v_o = v + v_r$
(d) Observer moves away from source at rest	$f_o = \left(\dfrac{v - v_r}{v}\right) f_e$	$\lambda_o = \lambda$	$v_o = v - v_r$
(e) Source and observer moving toward each other	$f_o = \left(\dfrac{v + v_r}{v - v_s}\right) f_e$	$\lambda_o = \left(\dfrac{v - v_s}{v}\right) \lambda$	$v_o = v + v_r$
(f) Source and observer moving apart	$f_o = \left(\dfrac{v - v_r}{v + v_s}\right) f_e$	$\lambda_o = \left(\dfrac{v + v_s}{v}\right) \lambda$	$v_o = v - v_r$

(Note: The subscript *o* refers to an *observed* quantity, *e* refers to an *emitted* quantity, *r* refers to the *receiver* (same as observer), *s* refers to the *source* and λ represents the wavelength had both the source and observer been stationary. Also, there is a difference between v_o, the observed velocity, and v_r, the velocity of the observer. The former is the velocity of the wave from the point of view of the observer, the latter is the velocity of the observer relative to the medium.)

(5) It's important to understand the mechanism whereby wave characteristics are determined. The emitted frequency of a wave is decided by the source. The nature of the medium plays no role in affecting this frequency, which is the same as the observed frequency when no motion of source and observer exists. The velocity of a wave, on the other hand, is decided by the medium and the source plays no role. For example, the speed of sound depends

on the temperature of the air through which it travels, the speed of a water wave depends on the depth of the body of water, the speed of a wave in a rope depends on the tension and density of the rope, and so on. The wavelength of a wave is determined by the other two decisions. Once the speed has been decided by the medium and the frequency by the source, the wavelength is *fait-accompli* - it has already been determined via the relationship $\lambda f = v$.

Many consequences flow from the above described decision making process. For example, the frequency of a reflected or refracted wave must be the same as that of the original wave since the source of the wave is the same. Once a wave enters a new medium, however, as in the case of refraction, its velocity (and consequently also its wavelength) is likely to change.

(6) The unit of amplitude depends on the type of wave. For a wave propagating in water or in a rope it is expressed in meters of height or depth. In the case of sound waves it is expressed in kg/m^3 (density) or N/m^2 (pressure). In the case of *EM* waves it is expressed in N/C (electric field intensity) or tesla (magnetic field intensity).

(7) The age-old debate between the wave and particle models of light has not been resolved in favor of either view. Light is too complicated an entity to be appropriately visualized as either wave or particle. It is, in fact, a subtle hybrid of both models, as depicted by the formulations of quantum mechanics. Some phenomena are best understood in terms of the particle view, because the particle characteristic of light becomes the dominant manifestation when the light experiences those phenomena. Other phenomena are best analyzed with the wave model in mind, because that characteristic of light dominates when the light experiences those phenomena. This

wave-particle duality extends also to items we are accustomed to thinking of as particles, such as electrons and protons - they sometimes behave as waves.

The wave model of light is used in analyzing interference, diffraction and polarization phenomena. The particle model of light is used in deciphering the details of the photoelectric effect, the Compton effect, and other phenomena. The basic rules of reflection and refraction (chapter fifteen) can be understood in terms of either model.

(8) In the photoelectric effect, it's important to recognize the difference between the number of ejected electrons (the current) and the energy of those electrons. The former is proportional to the intensity (brightness) of the *EM* radiation incident upon the photomaterial, and no *threshold* exists for this relationship (other than *zero*). The latter depends on the frequency of the *EM* radiation and a threshold does exists for this relationship - known as the *threshold frequency.*

(9) The popular dictum "light travels in straight lines" is grossly misleading. Light bends when passing through a single slit and when crossing a boundary into a different medium (refraction).

(10) In the formula $x/L = \lambda/d$ for the double slit phenomenon, the symbol x represents the distance between the center of the *central maximum* band of light and the center of the *first order* bright band (or, as a good approximation, the distance between the centers of any two adjacent bright lines). In the formula $y/L = \lambda/w$ for the single slit phenomenon, however, the symbol y represents the distance between the center of the central maximum and the center of the first order *dark* band.

(11) The unit for KE_{max}, W, E and V_0 in any of the formulae

above that contain these symbols must be the joule or the joule per coulomb. The electron-volt (eV) cannot be used as the unit for any of these quantities, unless the formulas are revised to accomodate such a shift. All quantities of energy presented in eV's must be converted to joules before they are plugged into any of the formulas presented above.

(12) In the formula $d\sin\theta = m\lambda$ for the bright bands produced by a double slit, the symbol m represents the band number, counting from the central maximum in either direction. The central maximum band itself is counted as *zero*, the first order band as *one*, the second order band as *two*, and so on. The angle θ in the formula is the angle illustrated below (figure 14-2) so long as it is relatively small (otherwise the formula is invalid).

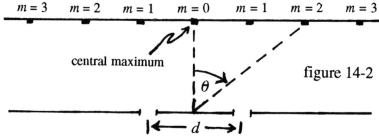

figure 14-2

An analogous system is employed when using the formula $W\sin\theta = m\lambda$ for the dark bands produced by a single slit. The first dark band on either side of the central maximum is labeled band number *one* ($m = 1$), the second is band number *two* ($m = 2$), and so on.

(13) In the formula $KE_m = hf - W$ the symbol f represents the incident frequency - the frequency of radiation that strikes the photo-material. This frequency may be equal to or greater than the threshold frequancy, f_o. It cannot be less than the threshold frequency.

(14) The formula $KE = mv^2/2$ (chapter twelve) cannot be applied

Waves and Particles

to photons. Instead, the mass-energy relationship for photons is provided by the formula $E = mc^2$ (chapter thirteen).

(15) The formula $p = h/\lambda$ functions as a two-way street. It provides the momentum of particles (such as photons) that are associated with waves of known wavelength and the wavelength of waves (the De Broglie wavelength) that are associated with particles of known momentum.

14.3 SOLVING THE PROBLEMS

(1) A tuning fork vibrates at the rate of 500 oscillations per second. The sound it creates travels in still air until it is reflected off a smooth wall 1200 meters from the tuning fork. The echo is heard by the person holding the tuning fork 8 seconds after the fork was set vibrating. (a) What is the wavelength of the sound wave? (b) How many compressions (pockets of compressed air) struck the wall during the eight second time interval?

SKETCH AND IDENTIFY

Every time a tuning fork swings to the right it creates a compression on its right side and a rarefaction (pocket of rarified, low density, air) on its left side; and every time it swings to the left it creates a compression on its left side and a rarefaction on its right side. These compressions and rarefactions then propagate through the medium, away from the tuning fork. The number of oscillations per second experienced by a tuning fork is therefore also the number of compression-rarefaction cycles emitted by the fork every second. The emitted frequency of the sound wave generated by our tuning fork is therefore 500 cycles/second or 500 Hz.

From the round trip distance of 2400 meters and the time it takes the sound wave to travel that distance, 8 seconds, we can compute the velocity of the wave (via the formula $vt = d$). Then the relationship $\lambda f = v$ can be employed to determine the wavelength.

The number of compressions striking the wall per second is the observed frequency of the wave - once the wave arrives at the wall. In this problem the observed frequency is equal to the emitted frequency, since no motion of source (tuning fork) or observer (wall) exists. But it takes the wave four seconds to reach the wall. Successive compressions strike the wall - at the rate of 500 times per second - only during the latter four seconds of the eight second interval between when the tuning fork was set vibrating and the echo was heard. Multiplying the frequency (500 Hz) by the time (four seconds) yields the total number of compressions that struck the wall during that time.

LINK UNKNOWN TO GIVEN

(a) $vt = d$
   $v(8) = 2400$ (round trip)
   $v = 300$ m/s
   $\lambda f = v$
   $\lambda(500) = 300$
   $\lambda = .6$ meters/cycle

(b) $N = ft$ (N is number of compressions to strike wall during four secs)
   $N = (500)(4)$
   $N = 2000$

(2) A tuning fork that oscillates 500 times per second is moving

Waves and Particles

away from you. Reaching your ear is a sound whose pitch corresponds to a frequency of 400 cps. At what speed is the tuning fork moving? (The speed of sound is 330 m/s at 0 °C, which you can assume is the case in this problem. Assume also that you are at rest relative to the medium, the air.)

SKETCH AND IDENTIFY

When a source of sound moves away from an observer, the observed frequency is less than the emitted frequency. The pitch heard by the observer in our problem, 400 cycles per second, is the observed frequency and the number of oscillations per second of the tuning fork (the source) is the emitted frequency. Use of the appropriate doppler effect formula should reveal the velocity of the source, v_s.

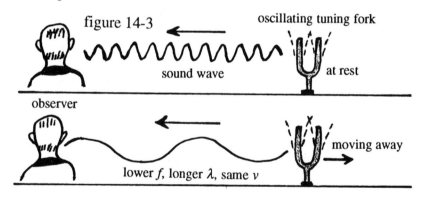

LINK UNKNOWN TO GIVEN

$f_o = [v/(v + v_s)]f_e$

$400 = [330/(330 + v_s)](500)$

$v_s = 82.5$ m/s

The tuning fork is moving at the rate of 82.5 m/s relative to the air. Since the observer is at rest relative to the air, that is also the speed of the tuning fork relative to the observer.

(3) A television transmitting antenna is located at T and a receiving antenna at R. An airplane is moving horizontally between these two locations, at some altitude above the level of both antennae (figure 14-4). The transmitting antenna emits *EM* waves of wavelength λ. The receiving antenna picks up two waves: one directly from the transmitter (path TR) and another by reflection off the airplane (path TAR). What relationship must exist between distances x, y and z (figure 14-4) and λ for (a) maximum constructive interference to occur at R, and (b) maximum destructive interference at R? Assume the law of reflection (chapter fifteen) is obeyed, meaning that x must equal y, and that the wave is inverted (crests become troughs and troughs become crests) upon reflection off the airplane.

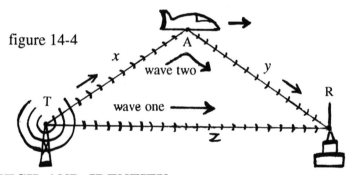

figure 14-4

SKETCH AND IDENTIFY

Wave *one*, which takes path TR, travels z meters as it goes from T to R. Wave *two*, which takes path TAR, travels $(x + y)$ meters in going from T to A to R. Since both rays begin their trips as part of the same expanding wave emanating from T, they both start in the same *phase* - at the time a crest emerges from T headed in the direction of TR, a crest also emerges from T headed along path TAR; and at the instant a trough embarks upon line TR, a trough also embarks upon path TAR. If the two path distances (z and $x + y$) are equal to each other or if they differ by a whole number of

Waves and Particles

wavelengths, a crest that arrives at R via path TR should meet a crest that took path TAR - so long as no inversion occurs upon reflection and the two waves travel at the same speed. But reflection does produce inversion in this case. Wave *two* experiences an inversion, but wave *one* does not. The effect of this inversion is to replace all the points on the reflected wave (A to R) that would otherwise have been crests with troughs, and all the points that would have been troughs with crests. Thus, when the two path distances are equal to each other or differ by a whole number of wavelengths, crests taking path TR actually meet up with troughs coming from A to R, and troughs arriving at R via path TR meet crests that arrive via path TAR. This situation results in maximum destructive interference of the two waves at R.

If the two path distances differ, not by a whole number of wavelengths, but by an odd number of half wavelengths ($\lambda/2$, $3\lambda/2$, $5\lambda/2$, ...) the two rays would arrive out-of-phase if there were no inversion upon reflection. Crests arriving at R via one path would meet troughs arriving via the other path, and vice-versa. But, again, inversion upon reflection replaces all crests with troughs and all troughs with crests, and this is experienced by only one of the rays (the reflected one). Thus, under these conditions, crests actually meet crests and troughs meet troughs. The two waves undergo maximum constructive interference at R.

LINK UNKNOWN TO GIVEN

For maximum destructive interference: difference in path distances is zero or a whole number of wavelengths.

$$(x + y) - z = m\lambda \qquad (m = 0, 1, 2, 3, ...)$$

For maximum constructive interference: difference in path

distances is an odd number of half wavelengths.

$$(x + y) - z = (m + 1/2)\lambda \qquad (m = 0, 1, 2, 3, ...)$$

Since $x = y$ we can simplify the above equations as follows:

(a) Maximum constructive interference: $2x = (m + 1/2)\lambda + z$

(b) Maximum destructive interference: $2x = m\lambda + z$

When x, z and λ are related as in equation (a), maximum constructive interference occurs at R and the antenna there receives the strongest possible signal. When these variables are related as in equation (b), maximum destructive interference occurs and the antenna at R receives no signal at all. The two waves completely destroy each other. When neither equation fits the three variables, an in-between situation results. Neither maximum construction, nor complete destruction, takes place.

(4) The stopping potential of a photomaterial is 0.36 volts when yellow light (5890Å) is incident upon it, and 3.14 volts when ultra-violet light (2537Å) is incident upon it. Find (a) the work function of the material and (b) the longest wavelength that ejects electrons from the material.

SKETCH AND IDENTIFY

Knowledge of the stopping potential, V_o, when a particular frequency of radiation is incident upon a particular photomaterial leads to knowledge of the maximum kinetic energy, KE_{max}, of the electons that are ejected from that photomaterial when that radiation strikes the material. The relationship between the two quantities, V_o and KE_m, is governed by the equation $KE_m = V_o e$. Once KE_m is ascertained the work function of the material, W, can be determined

Waves and Particles

from the formula $KE_m = hf - W$. Before using this equation, however, we will have to convert the given wavelength of the incident radiation into a frequency number. That can be accomplished via the relationship $\lambda f = v$.

Knowledge of the work function leads, in turn, to knowledge of the threshold frequency, f_0, via the formula $W = hf_0$. The threshold frequency has a corresponding threshold wavelength, λ_0, which can be determined from $\lambda_0 f_0 = v$. For frequency the threshold represents the minimum that ejects electrons; for wavelength the threshold represents the maximum that ejects electrons. (As frequency decreases, wavelength increases, since all *EM* waves have the same speed in vacuo - 3×10^8 m/s - and therefore also the same product of wavelength and frequency.)

Note: Different incident wavelengths and frequencies are associated with different electron kinetic energies and different stopping potentials even if the photomaterial is the same, but each photomaterial has only one work function, one threshold frequency and one threshold wavelength.

LINK UNKNOWN TO GIVEN

(a) From the data for yellow light:

$KE_{max} = V_0 e$

$KE_{max} = (.36)(1.6 \times 10^{-19}) = 5.76 \times 10^{-20}$ J

$\lambda f = v$

$(5.89 \times 10^{-7})(f) = 3 \times 10^8$ (1 Å = 10^{-10} meters)

$f = 5.1 \times 10^{14}$ Hz

$KE_{max} = hf - W$

$5.76 \times 10^{-20} = (6.63 \times 10^{-34})(5.1 \times 10^{14}) - W$

$$W = 33.8 \times 10^{-20} - 5.76 \times 10^{-20}$$
$$W = 2.8 \times 10^{-19} \text{ J}$$

Alternatively, from the data for ultra-violet light:
$$KE_{max} = V_o e$$
$$KE_{max} = (3.14)(1.6 \times 10^{-19}) = 5.02 \times 10^{-19} \text{ J}$$
$$\lambda f = v$$
$$(2.537 \times 10^{-7})(f) = 3 \times 10^8 \qquad (1 \text{ Å} = 10^{-10} \text{ meters})$$
$$f = 1.181 \times 10^{15} \text{ Hz}$$
$$KE_{max} = hf - W$$
$$5.02 \times 10^{-19} = (6.63 \times 10^{-34})(1.18 \times 10^{15}) - W$$
$$W = 7.82 \times 10^{-19} - 5.02 \times 10^{-19}$$
$$W = 2.8 \times 10^{-19} \text{ J}$$

(b) $\qquad W = hf_o$
$$2.8 \times 10^{-19} = (6.63 \times 10^{-34})(f_o)$$
$$f_o = 4.2 \times 10^{14} \text{ Hz}$$
$$\lambda_o f_o = v$$
$$\lambda_o (4.2 \times 10^{14}) = 3 \times 10^8$$
$$\lambda_o = 7.1 \times 10^{-7} \text{ m}$$

The longest wavelength that ejects electrons from this photomaterial is 7.1×10^{-7} meters, or 7100 Å.

(5) A 60-watt bulb radiates light whose average wavelength is 5000 Å. (a) How many photons does the bulb emit per second if 5% of the electrical energy supplied to it is converted to light? (b) How many of these photons impinge each second into a one square cm surface located 40 cm from the bulb? (Assume the surface is oriented perpendicularly to an imaginary line drawn to the bulb.)

SKETCH AND IDENTIFY

A power rating of sixty watts means that the bulb converts sixty joules of electrical energy into other forms of energy (light and heat) every second. Five percent of these sixty joules go into the production of photons, with every photon carrying an amount of energy provided by the formula $E = hf$. The average frequency, f, associated with each emitted photon can be determined from the average wavelength via the relationship $\lambda f = v$. The number of photons emitted per second, N, times the energy of each photon, E, should equal the amount of energy converted by the bulb to light every second, an amount equal to 5% of 60, or 3 joules.

The number of photons emitted by the bulb every second is the same as the number of photons that cross an imaginary spherical shell centered on the bulb, every second (figure 14-5). Since the

figure 14-5

process of photon emission is uniform in all directions, there is no reason to worry about the possibility that more photons are traveling, for example, to the right than to the left. In other words, the number of photons crossing any square cm of surface should be the same as the number crossing any other square cm on the same spherical shell, in the same amount of time. The surface area of an entire shell, A, is provided by the formula $A = 4\pi r^2$, where r is the radius of the shell. The ratio of the area of any segment of the shell, a, such as the one square cm in our problem, to the area of the entire shell, A, is equal to the ratio of the number of photons that cross that segment, x, to the number that cross the entire shell, N, every second. (This assumes, of course, that very many photons are emitted by the bulb every second, which is in fact the case.) In other words, $a/A = x/N$.

LINK UNKNOWN TO GIVEN

(a) $\quad \lambda f = v$

$(5 \times 10^{-7})(f) = 3 \times 10^8 \quad\quad (1 \text{ Å} = 10^{-10} \text{ meters})$

$f = 6 \times 10^{14}$ Hz

$E = hf$

$E = (6.63 \times 10^{-34})(6 \times 10^{14})$

$E = 3.98 \times 10^{-19}$ J $\quad$ (energy of each photon)

$3 = N(3.98 \times 10^{-19})$

$N = 7.54 \times 10^{18}$ photons emitted per second

(b) $\quad A = 4\pi r^2$

$A = (4)(3.14)(40)^2 \quad$ (area of sphere, 40 cm radius)

$A = 20{,}106$ cm^2

$a/A = x/N$

$1/20{,}106 = x/7.54 \times 10^{18}$

$x = 3.75 \times 10^{14}$

This is the number of photons that impinge into the one square cm surface, located forty cm from the bulb, per second.

(6) *Diffraction* is sometimes defined as "the spreading of light behind an obstacle" and sometimes as "the pattern produced when light passes through a single slit". Show that the two definitions are synonomous.

SKETCH AND IDENTIFY

The central maximum band of light that appears when light passes through a single slit is substantially brighter and wider than the other bright bands that appear on either side of the central maximum. When one looks at this pattern in ordinary situations (such as when a ray of light makes its way through an openning in a doorway) without carefully examining the boundaries of the light, only the central maximum is noticed. The neighboring fainter, narrower bands are typically not easy to discern. Now, the size of the central maximum (its width) is related to the size of the slit, the distance to the slit and the wavelength of the light via the formula $y/L = \lambda/w$ where y is one half the width of the central maximum. To ascertain what happens to the size of the light-ray we see coming through a single slit (the central maximum) as we look at different distances from the slit, all we need do is examine the formula to see how y changes as the distance from the slit, L, increases or decreases.

LINK UNKNOWN TO GIVEN

$y/L = \lambda/w$

$y = (\lambda/w)(L)$

If λ and w, the wavelength and slit size, are kept fixed, it is clear from the above formula that as L increases, so does y. In other words, the farther from the slit we look, the wider is the central maximum (figure 14-6). The central maximum band, in effect,

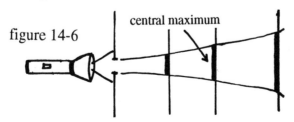

figure 14-6

"spreads out" as it travels away from the slit. Since both sides of the slit are "obstacles" to light (otherwise there is no "slit") and the central maximum band is essentially all we notice under ordinary circumstances, it appears as if "light spreads out behind an obstcale".

14.4 ON YOUR OWN PROBLEMS
(Asterisk indicates solution appears in appendices.)

*(1) Crests and troughs produced in a pool traverse the length of the pool in 4 seconds. The distance between successive crests is 20 cm and the length of the pool is 12 meters. (a) How many crests arrive at the end of the pool per minute? (b) How much time elapses between successive arrivals?

*(2) A railroad train traveling at 30 m/s emits a whistle sound of 600 cycles per second. What is the frequency and wavelength of the sound heard by (a) a stationary observer in front of the train and (b) a stationary observer behind the train? (The speed of sound in still air is 330 m/s.)

*(3) A ray of light of wavelength 5×10^{-7} meter is incident into a sheet of transparent material, as illustrated in figure 14-7. Some of the light is reflected off the front surface of the sheet (ray *one* in figure 14-7) while the rest is transmitted into the material. Some of this transmitted light is in turn reflected off the back surface of the

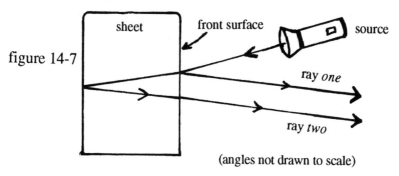

figure 14-7

(angles not drawn to scale)

Waves and Particles 319

sheet, passes through the front surface, and emerges from the sheet traveling in the same direction as, and overlapping, the ray that was reflected off the front surface (this is ray *two* in figure 14-7). (a) How thick should the sheet be in order that the two reflected waves interfere with each other constructively? (b) How thick must the sheet be for them to interfere destructively? Assume the incident ray strikes the sheet perpendicularly to its surface and that, as a result, both reflected rays travel perpendicularly to both surfaces of the sheet (which are parallel to each other). Assume also that the wavelength of the light is, for all practical purposes, the same inside the material as outside and take into account the fact that the first reflected wave is inverted upon reflection, while the second is not.

*(4) The threshold frequency of potassium is 5×10^{14} Hz. If a positively charged sheet of potassium is bombarded by *EM* radiation of wavelength 3600Å, (a) what minimum voltage is needed to turn back all the electrons ejected from the potassium before they reach an oppositely charged sheet situated parallel to the potassium sheet, (b) what is the velocity of the fastest moving electrons to come out of the potassium? (The mass of an electron is 9.1×10^{-31} kg.)

*(5) A beam of electrons is passed through two slits, producing an interference pattern on a phosphorescent screen located two meters from the slits. What is the distance between adjacent bright bands in the interference pattern (lines where the screen glows because electrons arrived there) if the slits are 4 mm apart and the velocity of the electrons is 2×10^6 m/s? (The mass of an electron is 9.1×10^{-31} kg.)

(6) A sonar wave (sound wave in water) has a frequency of 1500 Hz and a wavelength of one meter. (a) What is its speed? When the wave arrives at the surface of the water it crosses the

water-air boundary and is transmitted into the air. (b) What is its frequency now that it has become a sound wave in air? (c) What is its wavelength in the air? Assume the speed of sound in still air to be 330 m/s.

(7) Light of frequency 6×10^{14} Hz passes through a pair of slits that are 1.5×10^{-4} cm apart. What will be the separation between adjacent bright bands on a screen 0.6 meters from the slits?

(8) The threshold frequency of zinc is 9.7×10^{14} Hz. (a) What is the work function of zinc? (b) If zinc is irradiated with EM radiation of frequency 6.2×10^{15} Hz, what is the maximum kinetic energy of the electrons ejected from the zinc? (c) What is the stopping potential under these conditions?

(9) The work function of chromium is 4.6 eV. What is the threshold frequency of chromium ?

(10) What is the mass of a photon of yellow light of wavelength 5700Å?

(11) Light of wavelength 4.8×10^{-5} cm passes through a narrow slit and produces a diffraction pattern on a screen 150 cm from the slit. The distance between the center of the central maximum and the first order dark band is 0.4 cm. How wide is the slit?

(12) A single slit diffraction pattern is formed on a screen situated 2 meters from the slit. The slit is 0.1 mm wide. If the distance from the fourth dark band on one side of the central maximum to the fourth dark band on the other side is 10 mm, what is the wavelength of the light?

(13) A speaker generates sound by vibrating 500 times per second. If the speaker moves to the right at the rate of 40 m/s and an observer moves to the left, toward the speaker, at the rate of 20 m/s, what will be the frequency of the sound heard by the observer?

Waves and Particles

Assume the speed of sound in still air is 330 m/s.

(14) Compare the wavelength in problem thirteen to what the wavelength would have been had neither the speaker nor the observer been moving?

(15) Repeat problems thirteen and fourteen with the same respective velocities but now the observer and speaker are moving away from each other.

(16) What will be the change in stopping potential for electrons ejected from a particular photomaterial if the wavelength of incident radiation is decreased from 4500 Å to 4400Å?

(17) A rule of thumb for finding the distance to a lightning bolt is to begin counting when the flash is observed and stop when the thunder is heard. The number of seconds counted is then divided by three and the result so obtained is supposed to equal the distance to the lightning bolt in kilometers. Does this procedure (always, sometimes, never) yield the correct result? Explain.

(18) Two speakers separated by 8 meters emit sound waves of the same amplitude and frequency. The wave each speaker emits is *in phase* with the wave emitted by the other - compressions and rarefactions are generated simultaneously. A listener sits directly in front of one speaker at a distance of 6 meters, such that the two speakers and the listener form a right triangle. Find the three lowest frequencies for which the listener observes (a) maximum constructive interference and (b) maximum destructive interference.

(19) Assume the speakers in problem eighteen generate the lowest frequency sound for which the listener, located 6 meters directly in front of one speaker, observes maximum constructive interference. The listener then begins to walk in the direction parallel to the line connecting the speakers, bringing him closer to the other

speaker. How far does he walk before he observes total silence (maximum destructive interference)?

(20) The diameter of the pupil of the human eye is about 6 mm. How far from a 100-watt yellow light bulb should the eye be in order that one photon crosses the pupil per second? Assume the bulb converts 10% of the electrical energy it receives to light and that it emits only yellow light of wavelength 5900Å.

15

CHAPTER FIFTEEN

GEOMETRIC OPTICS

15.1 VOCABULARY AND FORMULAS

WORD	SYMBOL	UNIT
Angle of incidence	θ_i	degree
Angle of reflection	θ_r	degree
Normal	N	-------
Angle of refraction	θ_r	degree
Index of refraction	n	-------
Critical angle	θ_c	degree
Object distance	d_o	meter
Image distance	d_i	meter
Focal point	F	-------
Focal length	f	meter
Image size	S_i	meter
Object size	S_o	meter
Center of curvature	C	-------
Radius of curvature	R	meter

DEFINITIONS

NORMAL - Line drawn perpendicularly to a surface at the point where a ray of light strikes the surface.

ANGLE OF INCIDENCE - Angle between incident ray of light and normal.

ANGLE OF REFLECTION - Angle between reflected ray of light and normal.

ANGLE OF REFRACTION - Angle between refracted ray of light and normal.

INDEX OF REFRACTION - The ratio of the speed of light in vacuo (3×10^8 m/s, represented by the symbol c) to the speed of light in a particular material medium. This ratio is different for different material media (see chart, appendix D).

CRITICAL ANGLE - The angle of incidence for which the corresponding angle of refraction is 90°. This is possible only when the speed of light in the second medium is greater than in the first and, consequently, the light bends away from the normal.

OBJECT DISTANCE - Distance between source of light and lens or mirror, measured along the principal axis. (In the case of a plane mirror the distance is measured along a line drawn from the source perpendicularly to the mirror.)

IMAGE DISTANCE - Distance between image and lens or mirror, measured along the principal axis. (In the case of a plane mirror the distance is measured along a line drawn from the image perpendicularly to the mirror.)

FOCAL LENGTH - Distance between focal point and lens or mirror, measured along the principal axis.

PRINCIPAL AXIS - See figures 15-1 thru 15-4 below.

FOCAL POINT - See figures 15-1 thru 15-4 below.

Geometric Optics

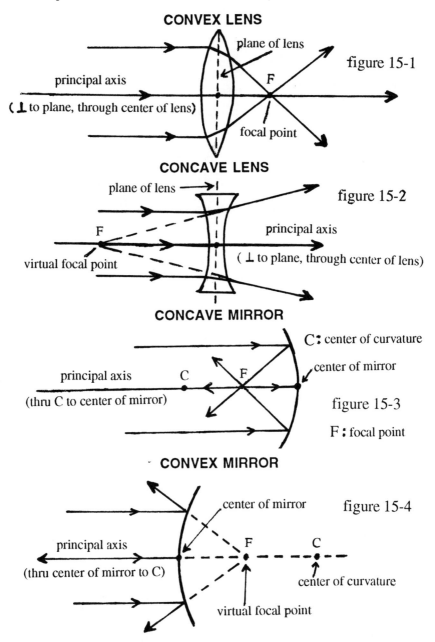

Note: The solid lines in figures 15-1 thru 15-4 above represent actual pencil-thin rays of light, whereas the broken lines are only imaginary extensions of those rays, drawn to indicate where the rays seem to be coming from.

CENTER OF CURVATURE - Point that constitutes the center of the sphere of which the concave or convex mirror is a segment.

RADIUS OF CURVATURE - Distance between the center of curvature and the mirror.

TOTAL INTERNAL REFLECTION - Rays of light that, upon encountering a boundary, do not cross the boundary into the next medium but are entirely reflected back into the medium from which they came. This occurs when the angle of incidence is greater than the critical angle. The reflected rays obey the law of reflection (Formulas, item *d*).

FORMULAS

a. $\sin\theta_i / \sin\theta_r = n_2/n_1$ **b.** $n = c/v$ **c.** $\sin\theta_c = n_2/n_1$
d. $\theta_i = \theta_r$ (reflection)   **e.** $1/d_o + 1/d_i = 1/f$ **f.** $S_i/S_o = d_i/d_o$
g. $R = 2f$ **h.** $d_i = d_o$ (plane surface)   **i.** $\sin\theta_c = 1/n$

15.2 THINGS TO KNOW

(1) The laws of reflection and refraction are applicable to individual pencil-thin rays of light emitted by a source. Light that spreads out from a source in all directions consists of a collection of many such rays closely spaced together (figure 15-5).

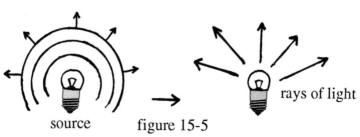

figure 15-5

Geometric Optics

(2) All angles of incidence, reflection and refraction in this chapter refer to the angle between the respective ray of light and the normal - not the surface.

(3) The index of refraction of a material may be different for different colors of light. If no color is specified, it is assumed that the light is yellow.

(4) Refraction (bending) of light does not occur inside a material but at the boundary between two media. For example, when a ray of light passes through glass it is refracted at the point of entry into the glass and again at the point of exit out of the glass. Inside the glass the ray travels in a straight line and no bending occurs (figure 15-6).

figure 15-6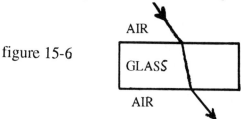

(5) When light crosses a boundary between two media and the speed of light is greater in the second medium, the light bends away from the normal (case A, figure 15-7). If, on the other hand, the speed of light is smaller in the second medium than in the first, the light bends toward the normal (case B, figure 15-7).

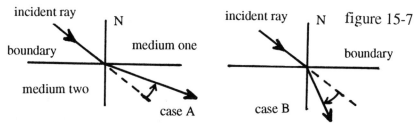

figure 15-7

(6) A ray of light that strikes a boundary perpendicularly is not refracted as it crosses that boundary. The angles of incidence and refraction are then both *zero* and the sine of each of these angles is

zero. When these values are plugged into the formula for refraction (Formulas, item *a*) we get *zero* divided by *zero*. This is mathematically neither illegal nor undefineable. The value of such a ratio is decided by the context in which it appears. In this case, it is made equal to the ratio of the sines for any other angle of incidence for that boundary (same pair and order of media and same color of light), namely the value of n_2/n_1.

(7) The formula $\sin\theta_c = 1/n$ is not valid for all media. It is applicable only to the case of light that goes from a material medium to empty space (or, as a good approximation, to air).

(8) Images are formed at meeting points. Real images are formed by rays of light that converge to a point; virtual images are formed by rays that diverge in such a manner that they can be extended backward to a point. The diverging rays then seem to the eye to be coming from that point. Rays that neither converge nor diverge, but are parallel, do not form an image of either type - they have no meeting point.

(9) Critical angles and total internal reflection can occur only at a boundary where the speed of light is greater in the second medium than in the first (and the index of refraction of the second medium is, consequently, smaller than that of the first).

(10) When constructing ray diagrams, locate the highest point in the object (the source of light) and draw two or, better yet, three rays from that point in the direction of the lens or mirror. The two or three rays should be selected from among those whose direction after encountering the lens or mirror is known without resorting to measuring angles. Such rays typically obey the following rules:

A. IN PARALLEL, OUT THROUGH (or from) *F*.

B. IN THROUGH (from or toward) *F*, OUT PARALLEL.

C. IN THROUGH CENTER, OUT STRAIGHT (lenses).

D. IN THROUGH (from or toward) *C*, OUT THROUGH (or from) *C* (mirrors).

> Note: These rules are based on the definitions of *focal point, center of curvature* and *principal axis* and, as such, are best understood and applied in conjunction with figures 15-1 thru 15-4. In keeping those figures in mind be aware of the fact that the arrows in those figures can be reversed and the diagrams remain correct. In using the rules the following code is applicable: The word *parallel* means parallel to the principal axis, *in* refers to the light as it approaches the lens or mirror, *out* refers to the light on its way away from the lens or mirror (after having been effected by the encounter), *center* refers to the center of a lens, *C* refers to the center of curvature of a mirror and *F* refers to the focal point of the lens or mirror. The rules are valid only for a lens that is thin and a mirror (concave or convex) that is a small segment of a sphere. (The same is true of formulas *e* and *f*).

The point where these rays meet (or seem to be coming from) after having been acted upon by the lens or mirror is the point where *all* the rays emanating from the chosen object point (the highest point in the object) meet (or seem to be coming from) after being acted upon by the lens or mirror. As such, this meeting point becomes the location of the image of the chosen object point. The orientation and size of the image of the entire object is then obtained by drawing a line from this image point to the principal axis (assuming the object is situated on the principal axis). The length of this line is the size of the image.

(11) Be aware of negative signs and their implications. The focal length, f, is negative for concave lenses and convex mirrors, the two instruments whose focal points are virtual (figures 15-2 and 15-4). A negative value for d_i indicates that the image is not on the same side of the lens or mirror as the light rays after they encounter the lens or mirror. Such images must be virtual.

(12) When the object distance, d_o, is infinitely large, the ratio $1/d_o$ becomes *zero*. The image then appears at the focal point, $d_i = f$, and the incoming rays are, for all practical purposes, parallel to each other and the principal axis. When the image distance, d_i, becomes infinitely large (the object is at the focal point) there is no image and the outgoing rays are parallel to each other.

(13) All other factors being equal, the larger a lens or mirror, the brighter the image, since more of the rays emanating from the source then encounter the lens or mirror and are directed to the image. The larger an image, on the other hand, the fainter it is.

(14) All real images are inverted; all virtual images are upright.

(15) A convex lens has two focal points, one on each side of the lens. A concave lens has two virtual focal points, one on each side of the lens. The focal lengths are equal on both sides of these lenses.

15.3 SOLVING THE PROBLEMS

(1) A ray of light is incident upon a 60°-60°-60° prism composed of crown glass, as illustrated (figure 15-8). Sketch the path of the ray all the way through and then out of the prism. The index of refraction of crown glass is 1.52.

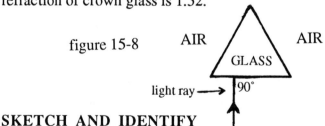

figure 15-8

SKETCH AND IDENTIFY

At every air-glass boundary the rules of optics are applied to ascertain the direction of the ray after it encounters the boundary.

Geometric Optics

The first question to address at every boundary is: will the ray go straight across the boundary, will it be refracted, or will it be internally reflected? If it is refracted, we then ask: What is the angle of refraction? The answers to these questions emerge from Snell's law (Formulas, item a) - $\sin\theta_i/\sin\theta_r = n_2/n_1$. If Snell's law yields no solution for θ_r ($\sin\theta_r > 1$) we conclude there is no angle of refraction, because no refraction takes place. The angle of incidence must be larger than the critical angle and total internal reflection occurs. The ray then obeys the law of reflection (Formulas, item d) - $\theta_i = \theta_r$. If Snell's law does yield a solution for θ_r, the value of θ_r reveals whether the ray goes straight through ($\theta_r = 0$), bends toward the normal ($\theta_r < \theta_i$) or away from the normal ($\theta_r > \theta_i$).

LINK UNKNOWN TO GIVEN

No refraction occurs at the first boundary since the ray strikes perpendicularly to the boundary (figure 15-8). An application of Snell's law to this boundary reveals that both θ_i and θ_r are equal to *zero* (Things to Know, item six). The light goes straight across the boundary, into the glass, and continues to travel in a straight line

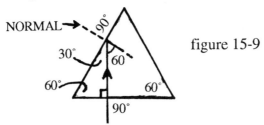

figure 15-9

until it encounters the next boundary, (figure 15-9). The angle between this boundary and the ray must be 30°, since the three angles of the triangle formed in the lower left corner of the prism must add up to 180°.

The angle of incidence at any boundary is, of course, the angle between the incident ray and the normal, not the boundary (see Definitions). So we draw a normal at the point where the ray strikes the boundary. The angle between the ray and this normal is 60°, since a normal, by definition, makes a 90° angle with the boundary. The angle of incidence, θ_i, at this boundary therefore is 60°.

We now apply Snell's law, keeping in mind that at this boundary the first medium is crown glass and the second is air.

$\sin\theta_i/\sin\theta_r = n_2/n_1$

$\sin 60°/\sin\theta_r = 1.00/1.52$ (n for air is 1.00 and for crown glass is 1.52)

$.866/\sin\theta_r = .66$

$\sin\theta_r = .866/.66 = 1.31$

There exists no angle whose sine is greater than 1.00. Snell's law breaks down this way whenever the angle of incidence is greater than the critical angle. To verify that 60° is greater than the critical angle for a crown-glass-to-air boundary, we proceed as follows:

$\sin\theta_c = n_2/n_1$

$\sin\theta_c = 1.00/1.52 = .66$

$\theta_c = 41°$

Sixty is indeed greater than forty-one. The net result is that the ray does not cross the boundary at all. Instead, it undergoes total internal reflection, obeying the law of reflection as it does so. The angle of reflection is, like the angle of incidence, 60° (15-10).

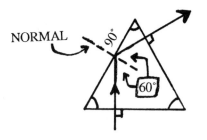

figure 15-10

Geometric Optics 333

After bouncing off this boundary the ray proceeds in a straight line through the glass until it encounters the next (and third) boundary. It strikes that boundary perpendicularly, since the three angles in the triangle formed in the upper corner of the prism must add up to 180° (figure 15-10). The ray then proceeds straight across this boundary without refraction, as illustrated, for reasons similar to its doing so upon encountering the first boundary.

(2) A ray of light is incident upon one side of an 8 cm cube of transparent plastic at an angle of 45° (figure 15-11). The index of refraction of the plastic is 1.6. By what distance is the ray that exits from the opposite side of the cube displaced from its original direction? In other words, what is distance x in figure 15-11?

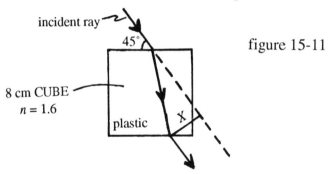

figure 15-11

SKETCH AND IDENTIFY

The first task is to draw a normal on the way into the plastic and on the way out, and determine all θ_i's and θ_r's. This should establish the angular relationship between the direction of the exiting ray and the original, incident ray (extended in figure 15-11 by the broken line). These two directions should turn out to be parallel to each other, otherwise no solution to our problem could possibly exists - a definite distance between two lines exists only if the lines are parallel to each other.

Once the angles of incidence and refraction are known we are in a position to calculate the displacement *x* via an application of basic trigonometry to the triangle formed by the refracted ray, line *x* drawn from the point where that ray meets the lower surface of the cube, and the segment of the broken line that completes the triangle (figure 15-11).

LINK UNKNOWN TO GIVEN

On the way into the plastic:

$\theta_i = 45°$ (given)

$\sin\theta_i / \sin\theta_r = n_2/n_1$

$\sin 45° / \sin\theta_r = 1.6/1.0$

$.707/\sin\theta_r = 1.6$

$\sin\theta_r = .707/1.6 = .44$

$\theta_r = 26°$

On the way out of the plastic:

$\theta_i = 26$

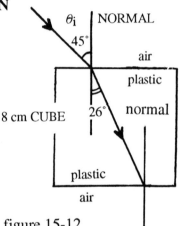

figure 15-12

(This is so since the problem states that the plastic is a *cube*. The upper surface is therefore parallel to the lower surface and normals to parallel surfaces (figures 15-12 and 15-13) must be parallel to each other. The angle of refraction on the way in and the angle of incidence on the way out are, consequently, *alternate interior angles* of a transversal. As such they must be equal to each other.)

$\sin\theta_i / \sin\theta_r = n_2/n_1$

$\sin 26° / \sin\theta_r = 1.00/1.6$

$.44/\sin\theta_r = .63$

$\sin\theta_r = .44/.63 = .70$

$\theta_r = 45°$

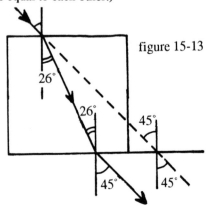

figure 15-13

Geometric Optics

Since the exiting ray and the broken line extension of the incoming ray both make 45° angles with the lower surface of the cube (extended in figure 15-13) they must be parallel to each other - as predicted earlier.

The distance x between two parallel lines is found by drawing a line perpendicular to both and determining the length of the segment between the lines. Let us draw this line segment from the point where the refracted ray meets the lower surface of the cube, as illustrated (figure 15-14). The angle between line x and the refracted ray, line y, is 71° (figure 15-14). Since lines x and y are constituent sides of a right triangle, knowledge of the length of y should lead to knowledge of the length of x via the relationship $\cos 71° = x/y$.

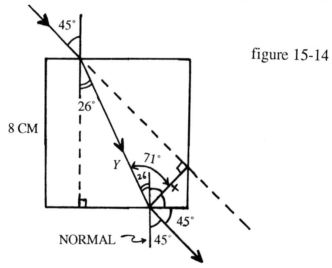

figure 15-14

To find the length of y let us extend the upper normal downward to the lower surface of the cube. This forms another right triangle with line y as hypotenuse and 26° as one acute angle (figure 15-14).

$\cos 26° = 8/y = .9$

$.9y = 8$

$y = 8.9$ cm

Now that y is known we return to the other right triangle and find x.

$\cos 71° = x/y$
$\cos 71° = x/8.9 = .33$
$x = 2.94$ cm

(3) A ray of light is incident upon the surface of a material medium at an angle of 30° from its surface. Part of the ray is reflected and part is refracted as it crosses the boundary into the material. The reflected and refracted rays are perpendicular to each other. (a) What is the index of refraction of the material? (b) What is the speed of light in the material? (c) If the color of the light is orange before it strikes the material, with a wavelength of 6 x 10^{-5} cm, what is the wavelength and color of the light as it travels inside the material?

SKETCH AND IDENTIFY

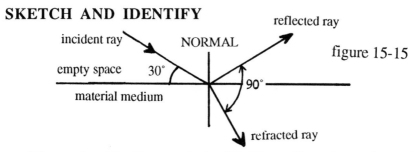

figure 15-15

The angle of incidence, θ_i, is 60°. The reflected ray obeys the law of reflection, and the refracted ray obeys the law of refraction. The angle between the normal and the reflected ray (θ_r) must, therefore, be equal to the angle between the normal and the incident ray (θ_i). Thus they both are 60°. From this and the provided fact that the reflected and refracted rays are perpendicular to each other we can readily ascertain the angle between the normal and the refracted

Geometric Optics

ray (figure 15-16). Applying Snell's law then leads to the index of refraction of the material.

Once the index of refraction, n, of the material is known, the relationship $n = c/v$ could be used to determine v, the speed of light in the material. The wavelength, λ, frequency, f, and speed, v, of any wave (such as light) in any medium are related to each other via the formula $\lambda f = v$ (chapter fourteen). Since the frequency is determined by the source of the wave, it doesn't change as the wave crosses a boundary from one medium into another. We therefore first apply $\lambda f = v$ to empty space (the first medium) to ascertain the light's frequency, then to the material (the second medium) to ascertain the wavelength and, upon consulting the appropriate chart in appendix D, the color of the light inside the material.

LINK UNKNOWN TO GIVEN

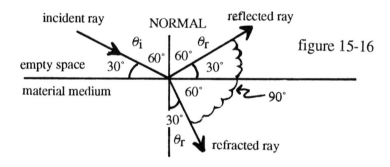

figure 15-16

(a) $\sin\theta_i/\sin\theta_r = n_2/n_1$

$\sin 60°/\sin 30° = n_2/1.00$

$.866/.5 = n_2$

$n_2 = 1.73$

(b) $n = c/v$

$1.73 = (3 \times 10^8)/v$ (c is speed of light in vacuo)

$v = 1.73 \times 10^8$ m/s

(c) In empty space   In the material
$$\lambda f = v$$ $$\lambda f = v$$
$$(6 \times 10^{-7})(f) = 3 \times 10^8$$ $$(\lambda)(5 \times 10^{14}) = 1.73 \times 10^8$$
$$f = 5 \times 10^{14} \text{ Hz}$$ $$\lambda = 3.46 \times 10^{-7} \text{ m}$$

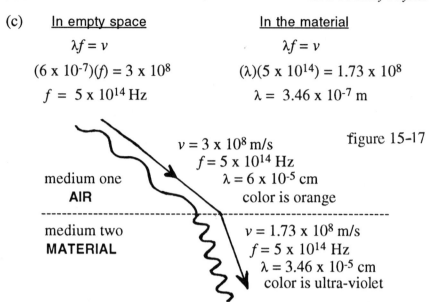

figure 15-17

(4) A 2 cm tall object is located 9 cm from a convex lens whose focal length is 6 cm. Find the location and size of the image using (a) the ray diagram method and (b) the formula method. Assume the object is situated on the principal axis.

SKETCH AND IDENTIFY

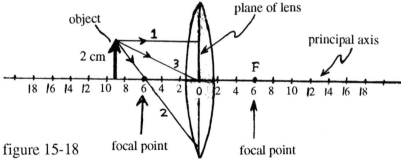

figure 15-18

all distances to center of lens

From the highest point of the object (the tip of the arrow) we draw three rays: one, parallel to the principal axis; two, through the focal point on the same side of the lens as the object; three, through

Geometric Optics 339

the center of the lens, as illustrated (figure 15-18). To each of these rays we apply the appropriate rule (Things to Know, item ten) to determine the direction taken by the ray after it encounters the lens. In using these rules we ignore the fact that each ray is in reality refracted twice, upon entering and exiting the lens. We simply extend each ray to the plane of the lens and proceed from there in the manner indicated by the applicable rule.

LINK UNKNOWN TO GIVEN

(a) Ray diagram method (figure 15-19 in conjunction with 15-1):

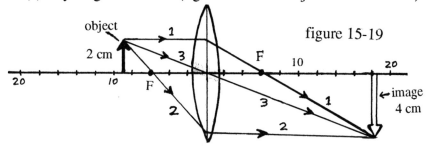

Ray *one*: **In** (to lens) **parallel** (to principal axis), **out** (from lens) **through** *F* (focal point on opposite side of lens).

Ray *two*: **In through** *F* (focal point on same side of lens), **out parallel** (to principal axis).

Ray *three*: **In through center** (of lens), **out straight**.

The point where these rays meet is the point where *all* the rays coming from the top of the object meet after passing through the lens. This point, therefore, constitutes the location of the image of the top of the object.

The complete image extends from this image-point to the principal axis. It is located approximately 18 cm from the lens, is 4 cm long and inverted. (Results obtained via ray diagrams are typically not as accurate as those obtained via the formulas.)

(b) Formula method:

$1/d_o + 1/d_i = 1/f$

$1/9 + 1/d_i = 1/6$

(cm may be used if all distances on both sides of the equation are so expressed.)

$1/d_i = 1/6 - 1/9$

$1/d_i = 3/18 - 2/18 = 1/18$

$d_i = 18$ cm

$S_i/S_o = d_i/d_o$

$S_i/2 = 18/9$ (cm may be used here for the same reason)

$S_i = (2)(18)/9 = 4$ cm

(5) The object in problem four is now moved to a point 3 cm distant from the convex lens. Find the location and size of the image via (a) the ray diagram method and (b) the formula method. Assume the object is still situated on the principal axis.

SKETCH AND IDENTIFY

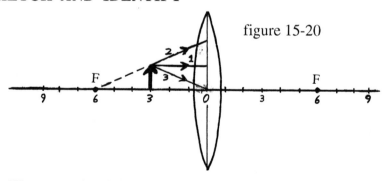

figure 15-20

We proceed as in problem four to draw three carefully chosen rays from the highest point of the object. But in this case ray *two* cannot run *through* the focal point, since the object is closer to the lens than the focal point. Any ray drawn from the object through the

Geometric Optics 341

focal point (on the same side of the lens as the object, of course) does not make it to the lens! Instead, ray *two* is drawn *from* the focal point, that is, in the direction coming from the focal point. This direction is determined via the broken line connecting *F* to the top of the object (figure 15-20). Then we apply the applicable rule (Things to Know, item ten) to each ray.

> (The game played here with ray *two* is not without ryme or reason. It is based on the fact that it matters not whether the ray of light actually passes through the focal point so long as it arrives at the lens coming from the direction of the focal point. Refraction is based on angles and direction, not distance traveled. Any ray coming from the direction of the focal point strikes the lens at such an angle that it emerges from the lens parallel to the principal axis.)

LINK UNKNOWN TO GIVEN

(a) Ray diagram method (figure 15-21 in conjuncton with 15-1):

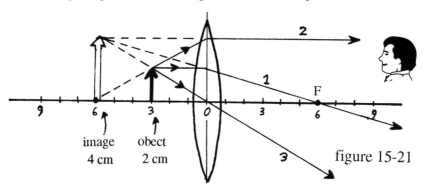

figure 15-21

Ray *one*: **In** (to lens) **parallel** (to principal axis), **out** (from lens) **through** *F* (focal point on opposite side of lens).

Ray *two*: **In from** *F* (focal point on same side of lens), **out parallel** (to principal axis).

Ray *three*: **In through center** (of lens), **out straight.**

These three rays do not meet anywhere! Instead, they diverge. This means that no real image is formed in this case. But the three rays seem to the eye to be coming from one point on the same side of the lens as the object. This can readily be ascertained by extending each of the rays backward, as illustrated by the broken lines in figure 15-21. This point constitutes the location of the virtual image of the top of the object. *All* the rays emanating from the top of the object are refracted in such a manner that they appear to the eye to come from this image point.

The complete image extends from this image-point to the principal axis. It is located approximately 6 cm from the lens on the same side of the lens as the object, is 4 cm tall and upright. Placing a screen at the image location does not, in this case, result in the appearance of an image on the screen (as is the case in problem four, where the image is real). But an eye on the right side of the lens in figure 15-21, looking leftward, intercepts the rays that passed through the lens and sees an enlarged, upright, virtual image at that location. You may recognize this arrangement as the way a magnifying glass may be used to read fine print.

(b) Formula method:

$1/d_o + 1/d_i = 1/f$

$1/3 + 1/d_i = 1/6$

$1/d_i = 1/6 - 1/3 = -1/6$

$d_i = -6$ cm

The negative value for d_i indicates that the image is virtual. It also implies, in the case of a lens, that the image appears on the same side of the lens as the object.

$S_i/S_o = d_i/d_o$

$S_i/2 = |-6|/3$ (When d_i is negative its absolute value is used here)

$S_i = (6)(2)/3 = 4$ cm

Geometric Optics

(6) A 2 cm tall object is situated 3 cm from a convex mirror whose focal length is 6 cm. Find the location and size of the image using (a) the ray diagram method and (b) the formula method. Assume the object is placed on the principal axis.

SKETCH AND IDENTIFY

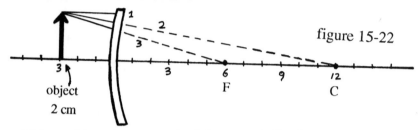

figure 15-22

This problem is approached in essentially the same manner as the previous two problems, except that some important nuances are different. In the case of a convex mirror there is no focal point on the same side of the mirror as the object. No ray can therefore be drawn through or from the focal point. Instead, we draw a ray (ray *three* in figure 15-22) *toward F*, that is, in the direction of the focal point on the other side of the mirror. This ray does not and cannot make it to that focal point - it is stopped and reflected by the mirror. The broken line extension of the ray beyond the mirror does, however, establish that the ray is headed straight toward the focal point.

Analogously, we draw a ray (ray *two* in figure 15-22) from the top of the object *toward C*, the center of curvature of the mirror. This point also is located on the other side of the mirror, at a distance twice as far from the mirror as the focal point.

Ray *one* (in figure 15-22) is the familiar *in parallel* type of ray. We are now ready to apply the appropriate rule to each of the three chosen rays.

344 How To Study Physics

LINK UNKNOWN TO GIVEN

(a) Ray diagram method (figure 15-23 in conjunction with 15-4):

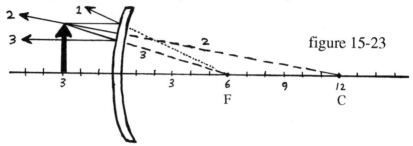

figure 15-23

Ray *one*: **In** (to mirror) **parallel** (to principal axis), **out** (from mirror) **from F** (focal point on opposite side of mirror).

Ray *two*: **In toward C** (center of curvature), **out from C**.

Ray *three*: **In toward F** (focal point on opposite side of mirror), **out parallel** (to principal axis).

The three rays diverge after being acted upon by the mirror. Consequently, they cannot form a real image. But to an eye that intercepts the reflected rays they seem to be coming from one point on the other side of the mirror (opposite the object). That point becomes the location of a virtual image of our chosen object point (the top of the object). All rays of light emanating from the top of the object that strike the mirror, seem to an eye that intercepts them to be coming from this image point.

To determine the location of the image point, simply extend each of the three reflected rays backward, to the right side of the mirror (figure 15-24). The image point turns out to be located approximately 2 cm from the mirror, on the side opposite the object. The image of the entire object is determined by drawing a line from this image point to the principal axis. The image of the entire object turns out to be one and one-third cm tall, smaller than the object. It is virtual and upright.

Geometric Optics

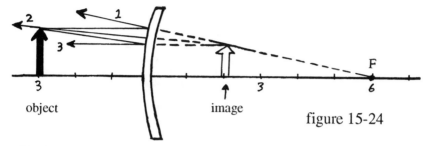

figure 15-24

(b) Formula method:

$1/d_o + 1/d_i = 1/f$

$1/3 + 1/d_i = 1/(-6)$ (*f* is negative - Things to Know, item eleven)

$1/d_i = -1/6 - 1/3$

$1/d_i = -1/6 - 2/6 = -3/6 = -1/2$

$d_i = -2$ cm

The negative value for d_i indicates that the image is virtual and not on the same side of the mirror (or lens) as the light rays after they encounter the mirror (or lens).

$S_i/S_o = d_i/d_o$

$S_i/2 = |-2|/3$

$S_i = 4/3$ cm

15.4 ON YOUR OWN PROBLEMS

(Asterisk indicates solution appears in appendices.)

*(1) What is the critical angle for light crossing a boundary from crown glass to water? (The index for crown glass is 1.52 and that of water is 1.33.)

*(2) At what angle from the surface does the ray in figure 15-25 exit from the prism, which is composed of crown glass?

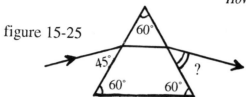

figure 15-25

*(3) What should be the angle of incidence for light going from air to crown glass in order that the angle of refraction is one-half the angle of incidence?

*(4) What must be the minimum length of a vertical plane mirror in order that a six foot tall person, standing 10 feet from the mirror, can see an image of his entire body? Does the answer depend on the distance between the person and the mirror?

*(5) A tree is situated 200 meters from a convex lens. On a screen one meter from the lens an 8 cm tall, inverted image of the tree appears. How tall is the tree?

*(6) A dentist uses a small concave mirror to obtain a good view of some hard-to-see places in her patient's mouth. If the radius of curvature of the mirror is 12 cm and the mirror is held 3 cm from a tooth, where will the image of the tooth appear? Will it be real or virtual? Upright or inverted? What is the ratio of the size of the image to the size of the tooth (a quantity known as *magnification*)? Obtain your answers via the ray diagram method.

(7) A 2 cm tall object is situated 9 cm from a concave lens whose focal length is 6 cm. Find the location and size of the image via (a) the ray diagram method and (b) the formula method. Assume the object is placed on the principal axis.

(8) A thick plate of glass with parallel surfaces is placed between a convex lens and its focal point. Will incoming rays parallel to the principal axis still converge at the focal point or will they converge at a point closer to, or farther from, the lens?

(9) A ray of light is incident upon a 45°-45°-90° crown glass prism, as illustrated in figure 15-26. Sketch the path of the ray all the way through and then out of the prism. (The index of crown glass is 1.42.)

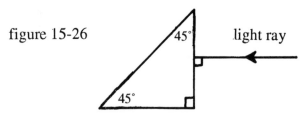

figure 15-26

(10) A point source of light is situated 3 cm from a smooth, horizontal surface. Sketch four rays emanating from the source in different directions, three of them striking the surface to the right of the source and one striking to the left. Draw the rays reflected off the surface, each in its appropriate direction. Then extend the reflected rays backward, below the surface. (a) Do the ray extensions meet at a point 3 cm below the surface, as they are supposed to ($d_i = d_o$)? (b) Is the object-image line perpendicular to the surface, as it is supposed to be?

(11) A ray of light is incident upon one side of a transparent object at an angle of 40° from the normal. The speed of light in the object is 1.4×10^8 m/s. (a) What is the angle between the ray and the surface it crossed, upon entering the object? (b) If the wavelength of the light before crossing the surface is 4.5×10^{-7} meters, what is its wavelength inside the object?

(12) A 3 cm tall object is located 12 cm from a convex lens whose focal length is 4 cm. Find the location, size and orientation of the image using (a) the ray diagram method and (b) the formula method. Assume the object is situated on the principal axis.

(13) A 4 cm tall object is located 10 cm from a concave mirror

whose focal length is 5 cm. Find the location, size and orientation of the image using (a) the ray diagram method and (b) the formula method. Assume the object is situated on the principal axis.

(14) A 2 cm tall object is located 6 cm from a convex lens whose focal length is 6 cm. Show that no image appears, either real or virtual, using (a) the ray diagram method and (b) the formula method. Assume the object is situated on the principal axis.

(15) The critical angle for light going from a material medium to air is 26°. What is the speed of light in the material?

(16) What is the angle of incidence for light going from air to diamond (index: 2.42) if the angle of refraction is 10°?

(17) Show that an object does not have to be directly in front of a plane mirror for an image of the object to appear in the mirror. Is the rule $d_i = d_o$ (Formulas, item h) applicable in such situations?

(18) An object is located 18 cm from a screen. (a) At what distances from the object can a convex lens of focal length 4 cm be placed in order to produce an image of the object on the screen? (b) What is the magnification of the image in each of these positions?

(19) A convex lens produces an image of an object on a screen situated 12 cm from the lens. When the lens is moved 2 cm farther from the object, the screen must be moved 2 cm closer to the lens in order to recreate the image on the screen. What is the focal length of the lens?

(20) As an object is moved closer to a convex lens, but not closer than the focal point, the distance between the lens and the image _____ and the size of the image _____ .

(21) How does the focal length of a convex lucite lens (index: 1.50) compare to the focal length of an identically shaped lens composed of flint glass (index: 1.61)?

APPENDIX A

SOLUTIONS

TO SELECTED *ON YOUR OWN* PROBLEMS

CHAPTER 2, MOTION

(1) 40 km/hr = 11.11 m/s

A. $v_f = v_i + at$
     $11.1 = 0 + a(8)$
     $a = 1.39$ m/s^2

B. $d = v_i t + at^2/2$
     $d = 0 + (1.39)(8)^2/2$
     $d = 44.48$ m

C. $v_f = v_i + at$
     $v_f = 0 + (1.39)(2)$
     $v_f = 2.78$ m/s

     $d = v_i t + at^2/2$
     $d = (2.78)(1) + (1.39)(1)^2/2$
     $d = 3.48$ m

(2) $d_1 = v_i t + at^2/2$ $d_2 = vt$ $v_f = v_i + at$
    $d_1 = 0 + (5)(12)^2/2$ $d_2 = (60)(100)$ $0 = 60 - 2t$
    $d_1 = 360$ m $d_2 = 6000$ m $t = 30$ sec
    $v_f = v_i + at$ $d_3 = v_i t + at^2/2$
    $v_f = 0 + (5)(12)$ $d_3 = (60)(30) + (-2)(30)^2/2$
    $v_f = 60$ m/s $d_3 = 900$ m
    $\bar{v} = d/t = (360+6000+900)/(12+100+30) = 51.13$ m/s

(3) A. $v_f^2 - v_i^2 = 2ad$
       $85^2 - 0 = 2(9.8)d$
       $d = 368.62$ m

B. $v_f = v_i + at$
     $85 = 0 + 9.8t$
     $t = 8.67$ sec

(4) 60 km/hr = 16.67 m/s

$v_f = v_i + at$ $\bar{v} = (v_i + v_f)/2$ $d = vt$
$0 = 16.67 - 12t$ $\bar{v} = (16.67+0)/2$ $d = (8.335)(1.39)$
$t = 1.39$ sec $\bar{v} = 8.335$ m/s $d = 11.59$ m

(5) A. $d_T = v_i t + at^2/2$ B. $x + 200 = d_c$ C. $v_f = v_i + at$
     $200 = 0 + (6)t^2/2$ $x + 200 = v_i t + at^2/2$ $v_f = (8)(8.165)$
     $t = 8.165$ sec $x + 200 = (8)(8.165)^2/2$ $v_f = 65.32$ m/s
                           $x = 66.67$ m

(6) $d = v_i t + at^2/2$ $d = v_i t + at^2/2$
      $400 = 0 + (9.8)t^2/2$ $400 = v_i (4.03) + (9.8)(4.03)^2/2$
      $t = 9.03$ sec $v_i = 79.51$ m/s
      $9.03 - 5.00 = 4.03$ sec

(7) $d = v_i t + at^2/2$ $d = v_s t_2$
      $d = 0 + (9.8)t_1^2/2$ $d = 330 t_2$
      $d = 4.9 t_1^2$
      $4.9 t_1^2 = 330 t_2$ $d = 70$ m
      $t_1 + t_2 = 4$
      $t_1 = 3.79$ sec, $t_2 = .21$ sec

CHAPTER 3, MOTION GRAPHS

(1) A. AB portion: $a = \Delta v/\Delta t = 0$ m/s^2, $d = vt = (20)(6) = 120$ m
    B. BC portion: $a = \Delta v/\Delta t = (50-20)/(12-6) = 5$ m/s^2,
       $d = v_i t + at^2/2 = (20)(6) + (5)(6)^2/2 = 210$ m
    C. CD portion: $a = \Delta v/\Delta t = (50-0)/(12-15) = -16.67$ m/s^2,
       $d = v_i t + at^2/2 = (50)(3) + (-16.67)(3)^2/2 = 75$ m

(2) A.

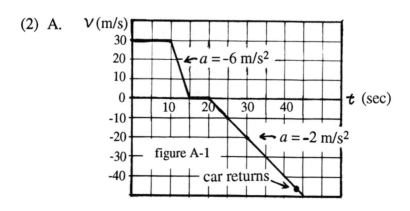

figure A-1

Solutions

B. Distance traveled moving away:
$(vt) + (v_i t + at^2/2) = (30 \cdot 10) + (30 \cdot 5 + [-6][5]^2/2) = 375$ m
All other intervals: velocity is either zero or negative.
Maximum distance = $150 + 375 = 525$ m

C. $d = v_i t + at^2/2 = 525$ (clock set to zero at $t = 20$)
$525 = 0 + (2)t^2/2$
$t = 22.9$ sec
The car returns at $t = 20 + 22.9 = 42.9$ sec

D.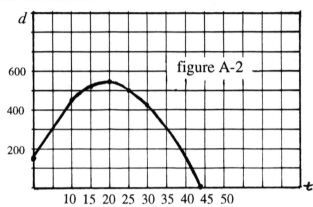

figure A-2

(3) A. AB: moving away
BC: not moving
CD: moving away

B. AB: velocity is positive
BC: zero velocity
CD: velocity is positive

C. AB: negative acceleration
BC: zero acceleration
CD: positive acceleration

D. For AB portion:
$\bar{v} = d/t = 50/20 = 2.5$ m/s
$\bar{v} = (v_i + v_f)/2$
$2.5 = (v_i + 0)/2$
$v_i = 5$ m/s
$a = \Delta v/\Delta t = (5-0)/(0-20) = -.25$ m/s²
For BC portion:
$a = \Delta v/\Delta t = (0-0)/(20-40) = 0$ m/s²
For CD portion:
$\bar{v} = d/t = (100-50)/20$ (setting $t = 0$ at point C)
$\bar{v} = 2.5$ m/s
$\bar{v} = (v_i + v_f)/2 = 2.5 = (0+v_f)/2$
$v_f = 5$ m/s
$a = \Delta v/\Delta t = (5-0)/(60-40) = .25$ m/s²

E.

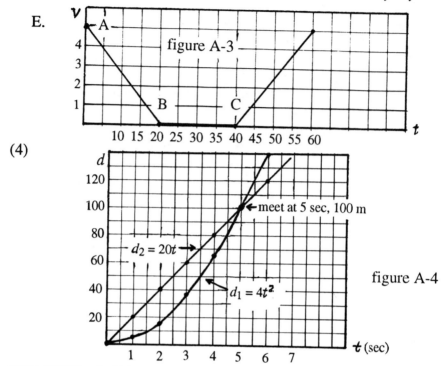

figure A-3

figure A-4

CHAPTER 4, VECTORS

(1) Scale: 5 km = 1 cm

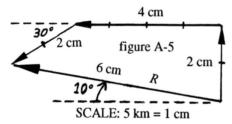

R: 6 cm = 30 km, 10° N of W

alternate solution

A = (0, 10) B = (-20, 0) C = (-8.7, -5)
R = (0-20-8.7, 10+0-5)
R = (-28.7, 5)

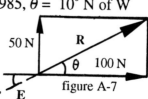

figure A-6

Magnitude of **R**: $R^2 = 5^2 + 28.7^2$, R = 29.14 km
Direction of **R**: $\cos\theta = 28.7/29.14 = .985$, $\theta = 10°$ N of W

(2) Resultant:
$R^2 = 50^2 + 100^2$, magnitude: 112 N

figure A-7

Solutions

$\cos\theta = 100/112 = .893$, direction: 27° N of E
Equilibrant:
magnitude: 112 N, direction: 27° S of W

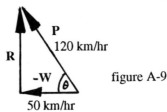
A-8

(3) $\tan 30° = 40/v_e$
$v_e = 40/\tan 30° = 69.3$ km/hr

(4) **R = W + P**
P = R - W
P = R + (-W)
$\cos\theta = 50/120 = .42$
Pilot should head 65° N of W
CHECK:

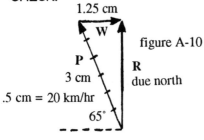

figure A-9

figure A-10

(5) $\cos 60° = F_x/200$
$F_x = (200)(.5) = 100$ N

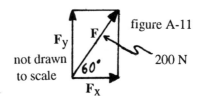

figure A-11

(6)

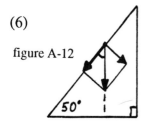

figure A-12

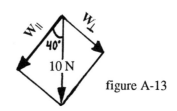

figure A-13

$W_{\parallel} = \cos 40°$, $W_{\parallel} = (10)\cos 40° = 7.66$ N
Frictional force must be 7.66 N strong.

(7)

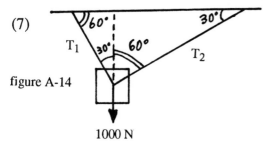

figure A-14

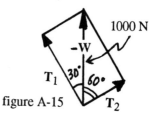

figure A-15

$T_1 + T_2 + W = 0$ $T_1/1000 = \cos 30°$ $T_1 = 866$ N
$T_1 + T_2 = -W$ $T_2/1000 = \cos 60°$ $T_2 = 500$ N

(8) $T_{1X} + T_{2X} + W_X = 0$
$T_{1Y} + T_{2Y} + W_Y = 0$
$-T_1 \cos 60° + T_2 \cos 30° + 0 = 0$
$T_1 \sin 60° + T_2 \sin 30° - W = 0$
$.87T_2 = .5T_1$
$W = .87T_1 + .5T_2$
$T_2 = (.5/.87)(T_1)$ $T_2 = (.575)(T_1)$
$W = .87T_1 + .288T_1 = 1.16T_1$

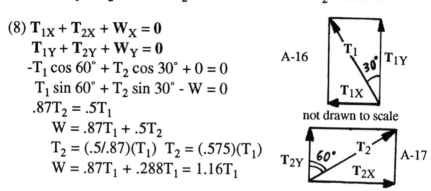

A-16

not drawn to scale

A-17

Setting T_1 at its maximum of 4000 N: $T_2 = 2300$ N, $W = 4640$ N (T_2 cannot be set at 4000 N because that would make T_1 equal to 6956 N - far above its allowed maximum.)
Maximum weight supportable therefore is 4640 N.

CHAPTER 5, LAWS OF MOTION

(1) 50 km/hr = 13.89 m/s, 90 km/hr = 25 m/s
$v_f = v_i + at$
$25 = 13.89 + a(20)$
$a = .55$ m/s^2
$F = ma$
$F = (5000)(.55) = 2750$ N

(2) $v_f^2 - v_i^2 = 2ad$
$0 - 100^2 = 2a(.12)$ (12 cm = .12 m)
$a = -41{,}666.67$ m/s^2
$F = ma$ (for bullet)
$F = (.010)(41{,}666.67)$ (10 gm = .010 kg)
$F = 416.67$ N
F(block) = F(bullet) = 416.67 N

(3) $F_{NET} = ma$
$W_1 - W_2 = (4+8)a$ (applied to whole system)
$m_1 g - m_2 g = 12a$
$(8)(9.8) - (4)(9.8) = 12a$
$a = 3.27$ m/s^2

Solutions

(4) $F_{NET} = ma$
$T - W = ma$
$T = ma + W$
$(20)a + (20)(9.8) \le 500$ $\qquad (W = mg)$
$20a \le 304$
$a \le 15.2 \text{ m/s}^2$

(5) $\qquad F_{NET} = ma$
a. $200 - (8)(9.8) - T = 8a$ $\qquad$ (upper mass)
b. $\qquad T - (5)(9.8) = 5a$ $\qquad$ (lower mass)
$T = 5a + 49$
$200 - 78.4 - (5a+49) = 8a$
$72.6 = 13a$
$a = 5.58 \text{ m/s}^2$
$T = (5)(5.58) + 49 = 76.9 \text{ N}$

alternate solution
$F_{NET} = ma$
$200 - (13)(9.8) = 13a$ $\qquad$ (whole system, ignore tension)
$72.6 = 13a$
$a = 5.58 \text{ m/s}^2$
$T - (5)(9.8) = (5)(5.58)$ $\qquad$ (lower mass)
$T = 76.9 \text{ N}$

(6) Scale reads tension in spring exerted upward on crate.
$F_{NET} = ma$
$T - W = ma$
a. $60 - m(9.8) = m(4)$ $\qquad (W = mg)$
$60 = (13.8)m$
$m = 4.35 \text{ kg}$
b. $\qquad W - T = ma$
$(4.35)(9.8) - 8 = (4.35)a$
$34.63 = (4.35)a$
$a = 7.96 \text{ m/s}^2$ $\qquad$ (downward)

The elevator is accelerating downward at the rate of 7.96 m/s² (either going down at increasing speed or going up at decreasing speed).

(7) $\cos 30° = W_{||}/W$
$W_{||} = W\cos 30°$ $\qquad$ figure A-18
$W_{||} = (m)(9.8)(.866)$
$W_{||} = 8.49(m)$

$F_{NET} = ma$
$W_\parallel = ma$
$8.49(m) = ma$
$a = 8.49 \text{ m/s}^2$
$v_f^2 - v_i^2 = 2ad$
$0 - 40^2 = 2(-8.49)d$
(going up incline: velocity is positive, force directed opposite to motion, speed decreases and a is negative.)
$-1600 = -16.98d$
$d = 94.23 \text{ m}$

(8) $F_{NET} = ma$
$F_p - W = (100,000)(5)$
$F_p - (100,000)(9.8) = (100,000)(5)$ $(W = mg)$
$F_p = 1,480,000 \text{ N}$

CHAPTER 6, FRICTION

(1) $F_{NET} = ma$
$\mu_k N = ma$
$\mu_k mg = ma$
$a = (.2)(9.8) = 1.96 \text{ m/s}^2$
$v_f = v_i + at$
$0 = 20 - 1.96t$
$t = 10.2 \text{ sec}$

(2) Slide begins when $W_\parallel = F_f$ (max)
$W\sin\theta = \mu_s N$ $N = W_\perp$
$mg\sin\theta = \mu_s mg\cos\theta$
$\tan\theta = \mu_s$
$\theta = \tan^{-1}(.3)$
$\theta = 17°$

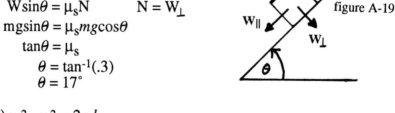

figure A-19

(3) $v_f^2 - v_i^2 = 2ad$
$0 - 30^2 = 2a(6)$
$a = -75 \text{ m/s}^2$
$F_{NET} = ma$
$\mu_k N = ma$

Solutions

$\mu_k mg = ma$
$(\mu_k)(9.8) = 75$
$\mu_k = 7.65$

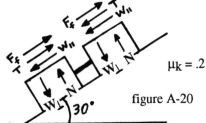

figure A-20

$\mu_k = .2$

(4) For 20 kg mass: $F_{NET} = ma$
$W_\| + T - F_f = ma$
(positive down incline, negative up incline)
$W\sin 30° + T - \mu_k W\cos 30° = ma$ ($N = W_\perp$)
$(20)(9.8)(.5) + T - (.2)(20)(9.8)(.866) = 20a$
$98 + T - 33.95 = 20a$
$T + 64 = 20a$ (equation one)
For 10 kg mass: $F_{NET} = ma$
$W_\| - T - F_f = ma$
$(10)(9.8)(.5) - T - (.2)(10)(9.8)(.866) = 10a$
$49 - T - 17 = 10a$
$32 - T = 10a$ (equation two)
Combining equations one and two: $T = 20a - 64$, $T = 32 - 10a$
$20a - 64 = 32 - 10a$
$a = 3.2$ m/s^2
$T = 0$ Newtons
(The rope is not stretched between the masses so it exerts no force of tension.)

CHAPTER 7, TWO-DIMENSIONAL MOTION

(1) $d_v = v_{iv}t + at^2/2$
$80 = (0)t + (9.8)t^2/2$
$t = 4.04$ sec
$d_H = v_H t$
$30 = v_H(4.04)$
$v_H = 7.4$ sec

(2) $v_{iH} = 200\cos\theta$, $v_{iv} = 200\sin\theta$
$v_{fv} = v_{iv} + at$ (upward trip)
$0 = 200\sin\theta - 9.8t$
$t_{up} = 200\sin\theta/9.8 = 20.4\sin\theta$
Total time in air $= (2)(20.4)\sin\theta = 40.8\sin\theta$
$d_H = v_H t$
$500 = (200\cos\theta)(40.8\sin\theta)$
$\sin\theta\cos\theta = .061$, $\theta = 3.5°$ and $86.5°$ (by trial and error)

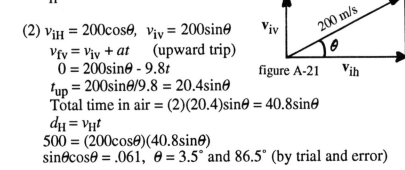

figure A-21

(3) 27.3 days = (27.3)(24)(60)(60) sec = 2.36 x 10⁶ sec
$a_c = 4\pi^2 r/T^2$

$$a_c = \frac{(4)(3.14)^2(3.6 \times 10^8)}{(2.36 \times 10^6)^2} = \frac{142 \times 10^8}{5.57 \times 10^{12}} = 2.55 \times 10^{-3} \text{ m/s}^2$$

(4) At $r = .04$ meters and $f = .55$ rev/sec the maximum force of static friction, F_f, is just equal to the centripetal force, F_c, necessary to maintain the circular track.
$F_c = F_f$
$4m\pi^2 r f^2 = \mu_s N$
$4m\pi^2 r f^2 = \mu_s mg$ (N = mg)
  $\mu_s = (4)(3.14)^2(.04)(.55)^2/9.8 = .049$
For $r > .04$ m: $4m\pi^2 r f^2 > \mu_s mg$ (not enough friction)
For $r < .04$ m: $4m\pi^2 r f^2 < \mu_s mg$ (more than enough friction)

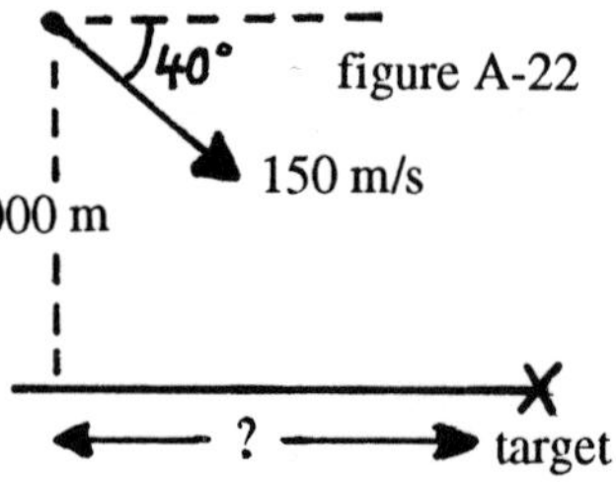

(5) $v_{iH} = 150\cos 40° = 115$ m/s
  $v_{iv} = 150\sin 40° = 96.4$ m/s
  $d_v = v_{iv}t + at^2/2$
  (velocity positive downward)
  $10{,}000 = 96.4t + (9.8)t^2/2$
  $0 = t^2 + 19.67t - 2040.8$
  $t = (-19.67 + \sqrt{386.9 + 8163.2'})/2$
  $t_1 = 36.4$ sec, $t_2 = -56.1$ sec (t_2 rejected)
  $d_H = v_H t = (115)(36.4) = 4{,}186$ m

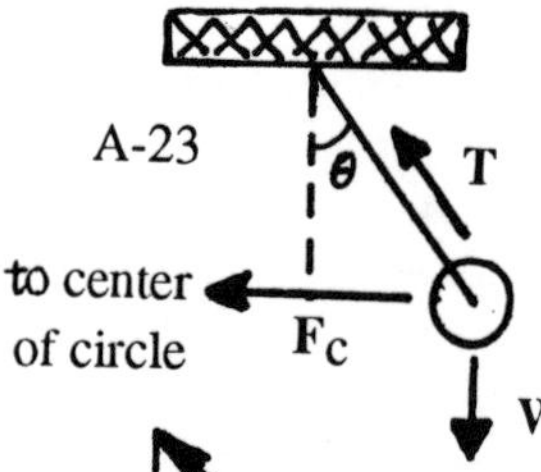

figure A-22

(6) $\mathbf{F}_{NET} = \mathbf{F}_c = \mathbf{T} + \mathbf{W}$
  $F_c = mv^2/r$
  $F_c = (m)(20)^2/300 = (1.33)(m)$
  $F_c/T = \sin\theta$ $(1.33)(m)/T = \sin\theta$
  $W/T = \cos\theta$ $mg/T = \cos\theta$
  $T = (1.33)(m)/\sin\theta$
  $T = mg/\cos\theta$
  $(1.33)(m)/\sin\theta = mg/\cos\theta$
    $\sin\theta/\cos\theta = 1.33/g = .136 = \tan\theta$
       $\theta = 7.7°$

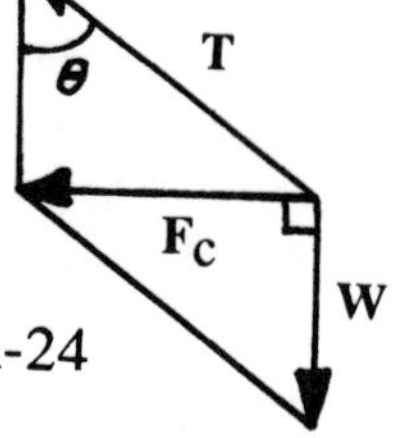

A-23

A-24

CHAPTER 8, GRAVITY

(1) Since $F \propto 1/d^2$, as d is tripled, d^2 becomes nine times as great

Solutions   359

and F is reduced to 1/9 its original value.
   New weight = (1/9)(200) = 22.2 N
           alternate method
$F_1 = GM_e m/d_1^2 = 200$ N $F_2 = GM_e m/d_2^2$ $d_2 = 3d_1$
$F_2 = GM_e m/(3d_1)^2 = (1/9)GM_e m/d_1^2 = (1/9)(200) = 22.2$ N

(2) $F_e = GM_e m/r_e^2 = 100$ N $F_u = GM_u m/r_u^2$
   $F_u = G(14.6M_e)m/(3.7r_e)^2 = (1.07)GM_e m/r_e^2 = 107$ N

(3) $T^2/a^3 = k$ (constant)
The moon and satellite revolve around the same body - the earth.
   $T_m^2/R_m^3 = T_s^2/R_s^3$ (R is radius of orbit, assumed circular)
   $(27.3)^2/(60R_e)^3 = (1)^2/R_s^3$ (T in days on both sides of equation)
   $(R_s^3)(745.3) = (2.16 \times 10^5)(R_e^3)$
           $R_s = (6.62)(R_e)$
R_s is the distance between the *center* of the earth (also the center of the orbit) and the satellite.
Distance from earth's *surface* to satellite is:
$6.62R_e - R_e = 5.62R_e = 3.6 \times 10^7$ m

(4) $W - N = F_c$
   Scale reading is *normal*, N = 0
      $W = F_c$
      $mg = 4m\pi^2 r/T^2$
      $9.83 = (4)(3.14)^2(6.4 \times 10^6)/T^2$
         $T = 5.07 \times 10^3$ sec, about 85 minutes

(5) $V_e = \sqrt{2GM/R}$
   $3 \times 10^8 \le \left[\dfrac{(2)(6.67 \times 10^{-11})M}{6.4 \times 10^6}\right]^{1/2}$
   $9 \times 10^{16} \le 2.08 \times 10^{-17} M$
      $M_e \ge 4.3 \times 10^{33}$ kg   (about 7×10^8 times its present mass)

CHAPTER 9, ELECTRICITY

(1) One mole = 6×10^{23}
   One gram of H_2O contains $(1/18)(6 \times 10^{23})$ or 3.33×10^{22} molecules.
   Each H_2O molecule consists of 10 electrons and 10 protons.

Total number of electrons: $(3.33 \times 10^{22})(10) = 3.33 \times 10^{23}$
Total numbers of protons: same as electrons
Total negative charge: $(3.33 \times 10^{23})(1.6 \times 10^{-19}) = 5.33 \times 10^4$ C
Total positive charge: same as negative charge
Radius of earth: 6.4×10^6 m
Distance between pole and equator, d, satisfies the relationship:
$d^2 = R^2 + R^2 = (6.4 \times 10^6)^2 + (6.4 \times 10^2)^2$
$d = 9.05 \times 10^6$ m
$F = kQ_1Q_2/d^2 = \dfrac{(9 \times 10^9)(5.33 \times 10^4)^2}{(9.05 \times 10^6)^2}$
$F = 3.12 \times 10^5$ N

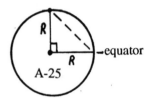

A-25

(2) $\mathbf{T} + \mathbf{F_e} + \mathbf{W} = 0$ (equilibrium)
$T_x + F_{ex} + W_x = 0$
$T_x = -F_{ex}$
$T_y + F_{ey} + W_y = 0$
$W_y = mg = -9.8 \times 10^{-3}$
$T_y = -W_y$
$T_y = 9.8 \times 10^{-3}$
$T_y/T_x = \tan 87°$ (magnitude only)
$9.8 \times 10^{-3}/T_x = 19.1$
$T_x = 5.13 \times 10^{-4}$ N
$F_e = 5.13 \times 10^{-4}$ N
$F = kQ_1Q_2/d^2$
$5.13 \times 10^{-4} = \dfrac{(9 \times 10^9)q^2}{(1 \times 10^{-1})^2}$
$q^2 = \dfrac{(5.13 \times 10^{-4})(10^{-2})}{9 \times 10^9}$
each $q = 2.39 \times 10^{-8}$ C

A-26

A-27 vertical

A-28

$\cos 87° = x/1$
$x = .05$ m
$d = x + x = .10$ m

(3)

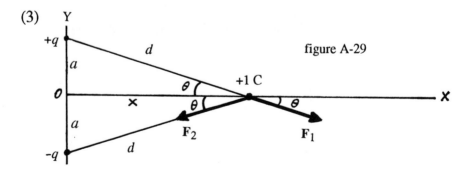

figure A-29

$F_1 = F_2 = kQ_1Q_2/d^2 = (9 \times 10^9)(q)(1)/(x^2 + a^2)$
$F_{NET} = F_1 + F_2$
F_{1x} is equal and opposite to F_{2x}.
F_{1y} and F_{2y} are both directed downward and are equal to each other.
$F_{1y} = F_1 \sin\theta \qquad F_{2y} = F_2 \sin\theta \qquad \sin\theta = a/d$
$F_{1y} = F_{2y} = \dfrac{(9 \times 10^9)q}{(x^2+a^2)} \cdot \dfrac{a}{d}$
$F_{NET, x} = F_{1x} + F_{2x} = 0$
$F_{NET, y} = F_{1y} + F_{2y} = (2)(9 \times 10^9)qa/(x^2+a^2)d$
The **F**-net vector is directed vertically downward (its x component is zero and its y component is directed downward). Its magnitude is equal to:
$F = (2)(9 \times 10^9)qa/(x^2+a^2)d$
As x grows larger and a remains fixed, $(x^2+a^2) \sim x^2$ and $d \sim x$.
$F_{NET} \sim (2)(9 \times 10^9)qa/(x^2)(x) \sim (1.8 \times 10^{10})qa/x^3$

(4) This is a two-dimensional motion problem. To find the time the electron spends between the plates: $v_H t = d_H$
$t = 6 \times 10^{-2}/3 \times 10^6 = 2 \times 10^{-8}$ sec
In the vertical dimension: $F = ma$ and $F = Eq$
$Eq = ma$
$(10^3)(1.6 \times 10^{-19}) = (9.1 \times 10^{-31})a$
$a = 1.76 \times 10^{14}$ m/s^2
$d_v = v_{iv}t + at^2/2$
$d_v = 0 + (.5)(1.76 \times 10^{14})(2 \times 10^{-8})^2 = 3.52 \times 10^{-2}$ m
$v_{fv} = v_{iv} + at$
$v_{fv} = 0 + (1.76 \times 10^{14})(2 \times 10^{-8}) = 3.52 \times 10^6$ m/s

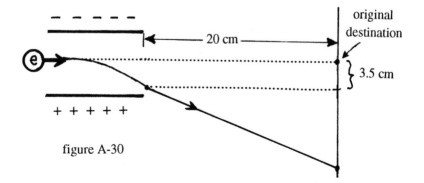

figure A-30

Once the electron emerges from the field its velocity vector continues unchanged (assuming gravitational effects can be ignored).
$\tan\theta = 3.52 \times 10^6 / 3 \times 10^6$
$\theta = 49.6°$
In the horizontal dimension: $v_H t = d_H$
$(3 \times 10^6)t = 2 \times 10^{-1}$ (20 cm = .2 m)
$t = 6.7 \times 10^{-8}$ sec
In the vertical dimension: $v_v t = d_v$
$(3.52 \times 10^6)(6.7 \times 10^{-8}) = d_v$
$d_v = 23.6 \times 10^{-2}$ meters = 23.6 cm

The electron strikes the screen 3.5 + 23.6, or 27.1, cm below its original destination.

CHAPTER 10, MAGNETISM

(1) Via hand rule one, using the left hand:

(2) A. Hand rule three, with the left hand:
   index finger: ←, thumb: X X X, middle finger: ↓
   Force is exerted downward: ↓
   B. $F = BIL$
   $F = (2 \times 10^4)(7.5)(.4)$
   $F = 6 \times 10^4$ N
   C. Density of flux lines = $B = 2 \times 10^4$ W/m²
   Area = 4×10^{-2} m²
   $\emptyset = BA = (2 \times 10^4)(4 \times 10^{-2}) = 800$ Webers

(3) $Bqv = F_c = 4m\pi^2 r f^2$
   $vT = d = 2\pi r$ (applied to each round trip at constant speed)
   $v(1/f) = 2\pi r$
   $v = 2\pi r f$
   $Bq(2\pi r f) = 4m\pi^2 r f^2$
   $f = Bq/2m\pi$
   $f = \dfrac{(6 \times 10^{-2})(1.6 \times 10^{-19})}{(2)(1.67 \times 10^{-27})(3.14)} = 9.15 \times 10^5$ Hz

(4) The electric force, F_e, is directed upward: ↑
   $F_e = Eq = (10^3)(1.6 \times 10^{-19}) = 1.6 \times 10^{-16}$ N
   If electrons are to go straight through: $\mathbf{F_{NET} = 0}$ (vertically)
   The magnetic force, F_m, must therefore be directed downward: ↓

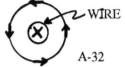
(test magnet in field)

$F_m = 1.6 \times 10^{-16}$ N
index finger (field intensity): X X X
thumb (moving charge): →
middle finger (force): ↓
$\quad F_m = F_e$
$\quad Bqv = Eq$
$\quad\quad v = E/B$
$4 \times 10^6 = 10^3/B$
$\quad\quad B = 2.5 \times 10^{-4}$ W/m²

(5) The magnetic field intensity at the S-pole is directed *toward* the bar magnet since that is the direction assumed by the S-to-N arrow (S→N) of a test magnet placed at that location.

figure A-35   figure A-36

According to hand rule two (left hand) such a field intensity is produced by negative charges that rotate counter-clockwise (as seen by someone who looks at the S-pole of the magnet). Bar magnets behave as they do because they are essentially solenoids (circular current). The electron rotations within a magnet are organized in such a manner that the field intensity created by each of them points in the same direction - toward the magnet at the S-pole, away from the magnet at the N-pole.

CHAPTER 11, MOMENTUM CONSERVATION

(1) $m_1v_1 + m_2v_2 = m_1v_1' + m_2v_2'$
$\quad\quad 0 = (200/9.8)v_1' + (10/9.8)(3)$ (W = mg)
$\quad 20.4v_1' = -3.06$
$\quad\quad v_1' = -.15$ m/s

(2) momentum before: $(.05)v_{bullet} + 0$ (kg·m/s)
momentum after: $(5.05)(10)$
$.05 v_{bullet} = 50.5$
$\quad v_{bullet} = 1010$ m/s

(3) In the horizontal dimension:
   Momentum before two minutes: (400)(20)
   Momentum after two minutes: (400+240)(v') (kg·m/s)
   (400)(20) = (640)(v')
       v' = 12.5 m/s
   (The vertical momentum of the raindrops is transferred to the whole earth.)

(4) A. $Ft = \Delta mv$
   The greater the stopping distance, d, the greater the stopping time, t. The greater t, on the other hand, the smaller the force, F, needed to accomplish the same Δmv - to eliminate all the momentum of the pitched ball. The smaller the force exerted by the hand on the ball, the smaller the force exerted by the ball on the hand.

B. $\bar{v} = (v_i + v_f)/2 = (40+0)/2 = 20$ m/s
    $d = \bar{v}t$
    $.50 = (20)t$ $t = .025$ sec
    $Ft = \Delta(mv)$
    $F(.025) = (40)(2) - 0$
    $F = 3,200$ N

(5)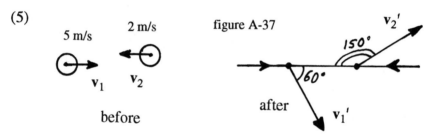

before   after

figure A-37

$m_1v_{1X} + m_2v_{2X} = m_1v_{1X}' + m_2v_{2X}'$
Since $m_1 = m_2$ all the m's drop out.
$(+5) + (-2) = v_1' \cos 60° + v_2' \cos 30°$
    $3 = .5v_1' + .866v_2'$
$m_1v_{1Y} + m_2v_{2Y} = m_1v_{1Y}' + m_2v_{2Y}'$
    $0 = (-v_1' \sin 60°) + v_2' \sin 30°$
      $.5v_2' = .866v_1'$
       $v_2' = 1.732v_1'$
      $3 = .5v_1' + (.866)(1.732v_1')$
   $v_1' = 1.5$ m/s and $v_2' = 2.6$ m/s

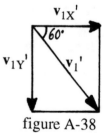

figure A-38

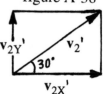

CHAPTER 12, FORMS OF ENERGY

(1) 500 watts = 500 joules of work done per second.
$$W = Fd\cos\theta$$
$$500 = F(5)\cos 30° \quad \text{(per second)}$$
$$F = 115.5 \text{ N}$$

(2) A. $V = Ed$
   $V = (1000)(.02)$ (2 cm = .02 m)
   $V = 20$ Volts
   $W = Vq$
   $W = (20)(1.6 \times 10^{-19})$ (charge on one proton is 1.6×10^{-19} C)
   $W = 3.2 \times 10^{-18}$ J

   B. The work done by the electric field = net work done, since the electric force is the net (and only) force.
   $$W_{NET} = \Delta KE$$
   $$3.2 \times 10^{-18} = \Delta KE = KE_f - KE_i = KE_f - 0$$
   $$KE_f = 3.2 \times 10^{-18} \text{ J}$$

(3) $V = IR$
   $120 = I(240)$
   $I = .5$ Amperes
   $P = VI$
   $P = (120)(.5)$
   $P = 60$ W or .06 kw
   kwhr = (kw)(hr)   (the *kwhr* is a unit of work)
   kwhr = (.06)(5) = .3 kwhr
   cost = (.3)(15) = 4.5 cents

(4) $Q = cm\Delta t$
   $Q = (1)(100)(100-30) = 7{,}000$ calories
   7,000 cal = 29,260 joules   (1 cal = 4.18 J)
   The minimum force needed to lift one kg is equal to the weight of the one kg object.
   $W = mg = (1)(9.8) = 9.8$ N
   $W = Fd\cos\theta$
   $29{,}260 = (9.8)(d)(1)$
   $d = 2{,}986$ meters, about 3 km!

(5) $\sin\theta = 16/80$
   $\theta = 11.5°$
   $W = mg = (10)(9.8) = 98$ N

figure A-41

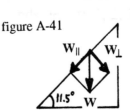

$\cos 11.5° = W_\perp/W = W_\perp/98 = .98$
$W_\perp = 96$ N
$\sin 11.5° = W_\parallel/W = W_\parallel/98 = .2$
$W_\parallel = 19.6$ N
$F_f = \mu_k N \qquad N = W_\perp$
$F_f = (.1)(96) = 9.6$ N

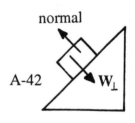

A-42

A. The work done against friction, W_f, is equal to the work done by the force that neutralizes friction - 9.6 N out of the 19.6 N exerted by gravity parallel to the incline.
$W_f = Fd\cos\theta$
$W_f = (9.6)(80)(1) = 768$ J

B. $W_{NET} = \Delta KE$
$W_{NET} = F_{NET}d\cos\theta$
$F_{NET} = 19.6 - 9.6 = 10$ N
$W_{NET} = (10)(80)(1) = 800$ J
$\Delta KE = KE_f - KE_i = KE_f - 0 = 800$ J
$KE_f = 800$ J

figure A-43

CHAPTER 13, CONSERVATION OF ENERGY

(1) $KE_1 + PE_1 = KE_2 + PE_2$
$mv_1^2/2 + mgh_1 = mv_2^2/2 + mgh_2$
$m(100)^2/2 + 0 = mv_2^2/2 + m(9.8)(200)$
$5000 = v_2^2/2 + 1960$
$v_2 = 78$ m/s

A-44

(2) Heat lost by metal = Heat gained by water
$c_m m_m \Delta t_m = c_w m_w \Delta t_w$
$c_m(50)(200-40) = (4.18)(100)(40-30)$
$c_m(8000) = 4180$
$c_m = .52$ J/gm·C°

(3) $KE_1 + PE_1 = KE_2 + PE_2 + $ heat
$mv_1^2/2 + mgh_1 = mv_2^2/2 + mgh_2 + Q$
$0 + (100)(9.8)(5) = (100)(3)^2/2 + 0 + Q$
$4900 = 450 + Q$
$Q = 4450$ J

A-45

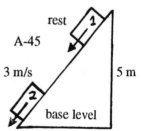

Solutions

(4) First, replacement of resistors in parallel:
$1/R_e = 1/60 + 1/60 + 1/60 = 3/60 = 1/20$
$R_e = 20\ \Omega$ (fig. A-46)
Second, replacement of resistors in series:
$R_e = R_1 + R_2 = 20 + 10 = 30\ \Omega$ (fig. A-47)
$V = IR$
$90 = I(30)$
$I = 3$ amperes
We now have the circuit in fig. A-48.
Next, determine voltage drops:
$V_4 = I_4 R_4$ $\qquad V_e = I_e R_e$
$V_4 = (3)(10) = 30$ V $\quad V_e = (3)(20) = 60$ V
Now, back to original circuit (fig. A-50):

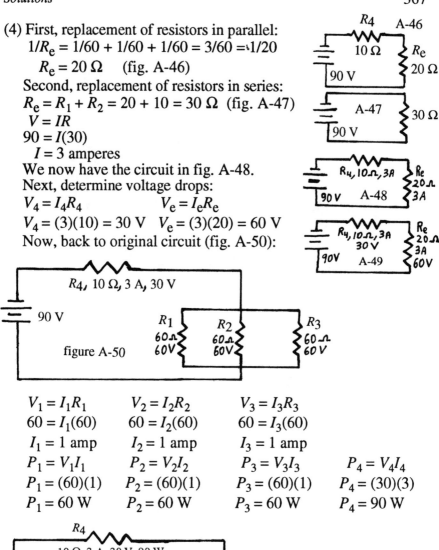

$V_1 = I_1 R_1$ $\qquad V_2 = I_2 R_2$ $\qquad V_3 = I_3 R_3$
$60 = I_1(60)$ $\qquad 60 = I_2(60)$ $\qquad 60 = I_3(60)$
$I_1 = 1$ amp $\qquad I_2 = 1$ amp $\qquad I_3 = 1$ amp
$P_1 = V_1 I_1$ $\qquad P_2 = V_2 I_2$ $\qquad P_3 = V_3 I_3$ $\qquad P_4 = V_4 I_4$
$P_1 = (60)(1)$ $\qquad P_2 = (60)(1)$ $\qquad P_3 = (60)(1)$ $\qquad P_4 = (30)(3)$
$P_1 = 60$ W $\qquad P_2 = 60$ W $\qquad P_3 = 60$ W $\qquad P_4 = 90$ W

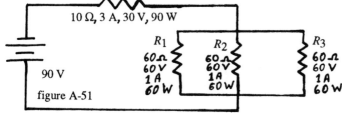

(5) A. Hand rule three, left hand:
  index finger (field): X X X
  thumb (moving electrons): →
  middle finger (force): ↓

Induced current is directed downward in straight wire segment (clockwise in closed loop).

$V = Blv$
$V = (8)(.1)(4) = 3.2$ volts
$V = IR$
$3.2 = I(.5) = 6.4$ amps

B. $F = BIL$
$F = (8)(6.4)(.1) = 5.12$ N (force exerted by field on current)
5.12 N of force is needed to oppose the force exerted by the field on the current.

C. $W = Fd\cos\theta$
To find d after one minute: $d = vt = (4)(60) = 240$ m
$W = (5.12)(240)(1) = 1228.8$ J

D. $P = VI = (3.2)(6.4) = 20.48$ W
$E = Pt = (20.48)(60) = 1228.8$ J

CHAPTER 14, WAVES AND PARTICLES

(1) A. $vt = d$
$v(4) = 12$
$v = 3$ m/s
$\lambda f = v$
$(.2)f = 3$ (20 cm = .2 m)
$f = 15$ Hz

15 crests strike the wall every second since the observed frequency is 15 cycles/sec.
$(15)(60) = 900$ (60 seconds = 1 minute)
900 crests strike the wall every minute.

B. $T = 1/f = 1/15 = .067$ sec

(2) In front of train, source moves *toward* observer at rest.
$f_o = [v/(v-v_s)]f_e$
$f_o = [330/(330-30)](600) = 660$ Hz
$\lambda_o = [(v-v_s)/v]\lambda$
$\lambda_o = [(330-30)/330]\lambda$
If train were stationary: $\lambda f_e = v$
$\lambda = v/f_e = 330/600 = .55$ m
$\lambda_o = (.91)(.55) = .50$ m

Behind train, source moves *away from* observer at rest.
$f_o = [v/(v+v_s)]f_e$
$f_o = [330/(330+30)](600) = 550$ Hz

$\lambda_o = [(v+v_s)/v]\lambda$
$\lambda = [(330+30)/330](.55) = .60$ m

	observed f	observed λ	observed v
Front	660 Hz	.50 m	330 m/s
Behind	550 Hz	.60 m	330 m/s

(3) Extra distance traveled by reflected ray *two* is $2w$ (round trip through sheet). Wave inversion occurs for ray *one*, reflected off front surface. No inversion occurs for reflected ray *two*.
A. Constructive interference:
$2w = m(\lambda/2)$ ($m = 1, 3, 5, ...$)
$w = m\lambda/4$
B. Destructive interference:
$2w = m\lambda$ ($m = 1, 2, 3, 4, ...$)
$w = m\lambda/2$

(4) A. $W = hf_o$
$W = (6.63 \times 10^{-34})(5 \times 10^{14}) = 3.315 \times 10^{-19}$ J
$\lambda f = c$
$(3.6 \times 10^{-7})f = 3 \times 10^8$
$f = 8.333 \times 10^{14}$ Hz
$KE_{max} = hf - W$
$KE_{max} = (6.63 \times 10^{-34})(8.33 \times 10^{14}) - (3.32 \times 10^{-19})$
$KE_{max} = 5.52 \times 10^{-19} - 3.32 \times 10^{-19}$
$KE_{max} = 2.2 \times 10^{-19}$ J
$KE_{max} = V_o e$
$2.2 \times 10^{-19} = V_o(1.6 \times 10^{-19})$
$V_o = 1.38$ volts
B. $KE_m = mv_m^2/2 = 2.2 \times 10^{-19}$
$(.5)(9.1 \times 10^{-31})v_m^2 = 2.2 \times 10^{-19}$
$v_m = 6.95 \times 10^5$ m/s

(5) $p = mv$
$p = (9.1 \times 10^{-31})(2 \times 10^6) = 1.82 \times 10^{-24}$ kg·m/s
$\lambda = h/p = 6.63 \times 10^{-34}/1.82 \times 10^{-24} = 3.64 \times 10^{-10}$ m
$x/L = \lambda/d$
$x = (3.64 \times 10^{-10})(2)/4 \times 10^{-3} = 1.82 \times 10^{-7}$ m

CHAPTER 15, GEOMETRIC OPTICS

(1) The formula $\sin\theta_c = 1/n$ cannot be used here because it is applicable only to situations where the second medium is empty space. Instead, we use $\sin\theta_c = n_2/n_1$:
$\sin\theta_c = 1.33/1.52 = .875$
$\theta_c = 61°$

(2) On the way in:
$\sin\theta_i/\sin\theta_r = n_2/n_1$
$\sin 45°/\sin\theta_r = 1.52/1.00$
$\sin\theta_r = .465$
$\theta_r = 28°$
angle A = 90 - 28 = 62°
angle B = 180 - 60 - 62 = 58°
θ_i on the way out = 90 - 58 = 32°
On the way out:
$\sin\theta_i/\sin\theta_r = n_2/n_1$
$\sin 32°/\sin\theta_r = 1.00/1.52 = .658$
$\sin\theta_r = .805$
$\theta_r = 54°$
angle C = 90 - 54 = 36°

figure A-52

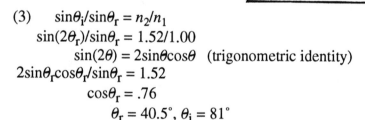

(3) $\sin\theta_i/\sin\theta_r = n_2/n_1$
$\sin(2\theta_r)/\sin\theta_r = 1.52/1.00$
$\sin(2\theta) = 2\sin\theta\cos\theta$ (trigonometric identity)
$2\sin\theta_r\cos\theta_r/\sin\theta_r = 1.52$
$\cos\theta_r = .76$
$\theta_r = 40.5°, \theta_i = 81°$

(4)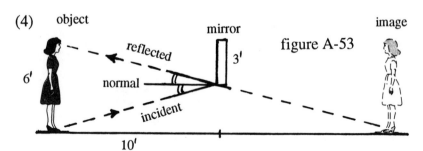

figure A-53

Answer: 3 ft. It does not depend on distance from mirror.

Solutions

(5) $S_i/S_o = d_i/d_o$
$.08/S_o = 1/200$ (8 cm = .08 m)
$S_o = 16$ meters   The tree is 16 meters tall.

(6) $R = 2f$ $12 = 2f$ $f = 6$ cm   Make object one cm tall.

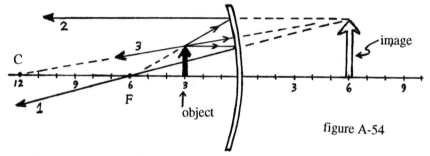

figure A-54

Ray *one*: in parallel, out through F.
Ray *two*: in from F, out parallel.
Ray *three*: in from C, out through C.
  The three rays diverge. No real image is formed. Instead, a 2 cm tall, virtual, upright image appears on the opposite side of the mirror, about 6 cm from the mirror ($d_i = -6$).
  Magnification $= S_i/S_o = 2/1 = d_i/d_o = |-6|/3 = 2$

APPENDIX B

LIST OF PHYSICAL CONSTANTS

Name	Value(s)
Gravitational constant	6.7×10^{-11} N·m²/kg²
Acceleration due to gravity	9.8 m/s²
Speed of light in a vacuum	3.0×10^8 m/s
Speed of sound at STP	3.3×10^2 m/s
Mass-energy relationship	1 u (amu) = 9.3×10^2 MeV
Mass of the Earth	6.0×10^{24} kg
Mass of the Moon	7.4×10^{22} kg
Mean radius of the Earth	6.4×10^6 m
Mean radius of the Moon	1.7×10^6 m
Mean distance from Earth to Moon	3.8×10^8 m
Electrostatic constant	9.0×10^9 N·m²/C²
Charge of the electron	1.6×10^{-19} C
One coulomb	6.3×10^{18} elementary charges
Electronvolt	1.6×10^{-19} J
Planck's constant	6.6×10^{-34} J·s
Rest mass of the electron	9.1×10^{-31} kg
Rest mass of the proton	1.7×10^{-27} kg
Rest mass of the neutron	1.7×10^{-27} kg

APPENDIX C
VALUES OF TRIGONOMETRIC FUNCTIONS

Angle	Sine	Cosine	Angle	Sine	Cosine
1°	.0175	.9998	46°	.7193	.6947
2°	.0349	.9994	47°	.7314	.6820
3°	.0523	.9986	48°	.7431	.6691
4°	.0698	.9976	49°	.7547	.6561
5°	.0872	.9962	50°	.7660	.6428
6°	.1045	.9945	51°	.7771	.6293
7°	.1219	.9925	52°	.7880	.6157
8°	.1392	.9903	53°	.7986	.6018
9°	.1564	.9877	54°	.8090	.5878
10°	.1736	.9848	55°	.8192	.5736
11°	.1908	.9816	56°	.8290	.5592
12°	.2079	.9781	57°	.8387	.5446
13°	.2250	.9744	58°	.8480	.5299
14°	.2419	.9703	59°	.8572	.5150
15°	.2588	.9659	60°	.8660	.5000
16°	.2756	.9613	61°	.8746	.4848
17°	.2924	.9563	62°	.8829	.4695
18°	.3090	.9511	63°	.8910	.4540
19°	.3256	.9455	64°	.8988	.4384
20°	.3420	.9397	65°	.9063	.4226
21°	.3584	.9336	66°	.9135	.4067
22°	.3746	.9272	67°	.9205	.3907
23°	.3907	.9205	68°	.9272	.3746
24°	.4067	.9135	69°	.9336	.3584
25°	.4226	.9063	70°	.9397	.3420
26°	.4384	.8988	71°	.9455	.3256
27°	.4540	.8910	72°	.9511	.3090
28°	.4695	.8829	73°	.9563	.2924
29°	.4848	.8746	74°	.9613	.2756
30°	.5000	.8660	75°	.9659	.2588
31°	.5150	.8572	76°	.9703	.2419
32°	.5299	.8480	77°	.9744	.2250
33°	.5446	.8387	78°	.9781	.2079
34°	.5592	.8290	79°	.9816	.1908
35°	.5736	.8192	80°	.9848	.1736
36°	.5878	.8090	81°	.9877	.1564
37°	.6018	.7986	82°	.9903	.1392
38°	.6157	.7880	83°	.9925	.1219
39°	.6293	.7771	84°	.9945	.1045
40°	.6428	.7660	85°	.9962	.0872
41°	.6561	.7547	86°	.9976	.0698
42°	.6691	.7431	87°	.9986	.0523
43°	.6820	.7314	88°	.9994	.0349
44°	.6947	.7193	89°	.9998	.0175
45°	.7071	.7071	90°	1.0000	.0000

APPENDIX D

INDICES OF REFRACTION

Air 1.00
Alcohol 1.36
Canada Balsam 1.53
Corn Oil 1.47
Diamond 2.42
Glass, Crown 1.52
Glass, Flint 1.61
Glycerol 1.47
Lucite 1.50
Quartz, Fused 1.46
Water 1.33

WAVELENGTHS OF LIGHT IN VACUO

($\lambda = 5.9 \times 10^{-7}$ m)

Violet .... 4.0 – 4.2 $\times 10^{-7}$ m
Blue .... 4.2 – 4.9 $\times 10^{-7}$ m
Green .... 4.9 – 5.7 $\times 10^{-7}$ m
Yellow .... 5.7 – 5.9 $\times 10^{-7}$ m
Orange .... 5.9 – 6.5 $\times 10^{-7}$ m
Red .... 6.5 – 7.0 $\times 10^{-7}$ m

SPECIFIC HEATS (kJ/kg·C°)

Alcohol (ethyl)	2.43 (liq.)	Platinum	0.13 (sol.)
Aluminum	0.90 (sol.)	Silver	0.24 (sol.)
Ammonia	4.71 (liq.)	Tungsten	0.13 (sol.)
Copper	0.39 (sol.)	Water { ice	2.05 (sol.)
Iron	0.45 (sol.)	water	4.19 (liq.)
Lead	0.13 (sol.)	steam	2.01 (gas)
Mercury	0.14 (liq.)	Zinc	0.39 (sol.)

INDEX

Accelerated motion,
  39-44
  distance traveled
    during, 37, 41-44
  graphs of, 55, 59, 64-73
Acceleration
  centripetal, 166-167
  on the moon, 185-186
  rate of, 5-6, 38, 39
Ampere, 250
Amplitude, 300, 304

Back EMF, 268, 271
Base level, of PE,
  252, 269
Boundary behavior, of
  wave, 327-328, 330

Center of curvature, 326
Central max., 305, 317
Centripetal acceleration,
  166-167
Centripetal force, 157,
  159
Charge(s)
  conservation of, 270
  moving, 214, 216
  point, 196
  unit of, 195
Charged particles
  in electric fields, 207,
    227-228, 264
  in magnetic fields,
    220-223
Circuits(s)
  compound, 279-284
  parallel, 279-281
  series, 276-279
Closed system, 234
Coherent waves, 310
Collisions, 236-243
Components, 80, 234
Compressions, 307

Concave lens, 346
Concave mirror, 346
Concurrent vectors, 81
Conservation
  charge, 270
  energy, 268-298
  internal energy,
    270-271, 288-290
  mass-energy, 272
  momentum, 233-247
Conservative forces, 252
Converging lenses,
  338-342
Converging mirrors, 346
Conversions, units, 15
Convex lens, 336-338
Convex mirror, 343-345
Coulomb, 195
Coulomb's law, 196
Critical angle, 324, 328
Current(s), electric, 210,
  250, 270
  action of currents on,
    216-220
  action of magnets on,
    213, 215-216
  action of, on magnets,
    212-213
  induced, 223-224,
    269, 285-288, 290

DeBroglie wave, 307
Derived units, 12-14
Diffraction, 316-318
Dimensional analysis, 14
Displacement, 81, 111
Diverging lenses, 346
Diverging mirrors,
  343-345

Doppler effect, 302, 308
Double-slit exp, 305

Duality
  (wave-particle), 304

Effective resistance,
  269, 277, 279-282
Elastic $P.E.$, 262-263
Electric field(s), 196-197
  diagrams of, 195, 197
  intensity of, 194,
    199-204
Electricity, 194-209
Electric power and
  energy, 258-260, 284
Electromagnetic waves,
  300, 305, 313
Electromotive force
  (EMF)
  back, 268, 271-272
  induced, 268
Electron(s)
  in electric field, 207,
    227
  thru two slits, 319
Electron volt, 306
Energy, 248-267
  conservation of,
    268-298
  elastic potential, 262
  electric power and,
    258-260
  heat, 260, 279
  internal, 288-290
  kinetic, 249, 273-274
  mass and, 272
  mechanical, 268
  nuclear, 297
  potential, see P.E.
  work and, 251-252
Equations in physics, 14
Equilibrant, 81, 92
Equilibrium, 81, 91,
  105-111
  thermal, 204-206

375

Escape velocity, 178, 188-190

Field(s)
  changing, 290-292
  electric, see Electric field(s)
  gravitational, 178
  magnetic, see Magnetic field(s)
Flux density, 211
Flux lines, 211
Focal length, 324-325
Focal point, 324-325
  virtual, 325, 329
Force, 6-7, 81, 116
  centrifugal, 157-158
  centripetal, 157, 159
  concurrent, addition of, 87-92, 102-111
  conservative, 252-253
  electric, 203-206
  friction, 141-156
  gravitational, 178-193
  magnetic, 211, 214-220
  net, 117
  normal, 97-98, 129-131
  resolution of, 99-102
  response, 117, 149, 151
  tension, 97, 125-128
Formulas, 7-10
Free fall, 43-44
  graphs of, 61-63
  on the moon, 185-186
Frequency
  circular motion, 158
  of wave, 300, 302
  threshold, 301
Friction, 141-156
  coefficient of, 141
  kinetic, 141, 144-148
  static, 141, 148-153, 171

Fundamental units, 12

Geometric optics, 323-349
Graphs
  accelerated motion, 55-56, 59-60, 64-73
  area under v-t, 70-71
  constant velocity, 54
  direct variation, 30-31
  free fall, 61-63
  inverse variation, 32-35,
  slope of, 59-61
Gravity, 178-193
Gravitational PE, 256, 273

Hand rules, 212-214
Head-to-tail method, 82, 87
Heat, 249
  cons. of, 270, 288
  internal energy and, 249, 288-289
  specific, 249
Hertz, 299
Hooke's law, 263

Image(s)
  concave lenses, 346
  concave mirrors, 346
  convex lenses, 338-342
  convex mirrors, 343-345
  real, 328, 330, 338-340
  size & location, 338-345
  virtual, 328, 330, 341-342, 344-345
Image distance, 324
Impulse, 233, 243-244
Incidence, angle of, 324

Index of refraction, 324
Induced current, 223, 269, 285-288, 290-292
Induced *EMF*, 268
Inertia, 115-116
Instant velocity, 61
Intensity
  electric field, 194, 199-204
  magnetic field, 210-211, 216
Interference, 300, 305, 310
Internal energy, 249
  cons. of, 288-290
  heat and, 249, 288
Internal reflection, 326
Inverse variation, 31-33
Inverse square variation, 33-35, 183

Joule, 249

Kepler's laws, 180
Kinetic energy, 249, 273
Kirchoff's laws, 270

Law(s)
  cons. of charge, 270
  cons. of energy, 268-298
  cons. of momentum, 233-247
  Coulomb's, 196
  Hooke's, 263
  Kepler's, 180
  Kirchoff's, 270
  Lenz's, 268, 271, 291
  Ohm's, 271, 277, 283
  of inertia, 115, 122
  of motion, 115-140
  of reflection, 326-327
  of refraction, 326-328, 330-338
  of universal gravity, 178-193

Index

Snell's, 326-328, 330-338
Lenses
  concave, 346
  converging, 338-342
  convex, 338-342
  diverging, 346
Lenz's law, 268, 271, 291
Light
  color and wavelength of, 336-338
  dual nature of, 304
  speed of, 336-338
  spreading of, 316-318
  wave nature of, 304

Magnets
  action of currents on, 211
  action of magnets on, 214, 216, 224
  action of, on currents, 213
  action of, on moving wires, 285-288
  natural, 214, 228
Magnetic field(s), 210
  action of, on currents, 213
  action of, on moving wires, 285-288
  action of, on magnets, 214, 216, 224
  changing, creating **E** fields, 290-292
  charged particles in, 220-223
  diagrams of, 214, 219
  flux density of, 211, 225-226
  induction of, 211
  intensity of, 210-211, 216

Magnetic force, 211, 214-220
Magnetism, 210-232
Magnification, 340, 342
Mass, 116, 118
  of earth, 179
  energy and, 272
Matter waves, 307
Measurement, 10-16
Metric system, 12
Mirrors
  concave, 346
  converging, 346
  convex, 343-345
  diverging, 343-345
  plane, 346
Momentum, 233, 238
  cons. of, 233-247
  in two dimensions, 234-235, 239-242
  photon, 304, 314-316
Motion, 36-53
  accelerated, 39-44
  circular, 166-169, 171
  graphs of, 54-79
  laws of, 115-140
  of charged particles, 214, 216, 207, 220-223, 227, 264
  orbital, 178, 180, 184
  two-dimensional, 157-177
  uniform, 37, 39, 45, 47
Newton, 116
Normal force, 97-98, 129-131, 141, 187
Normal line, 324
Nuclear energy, 297

Object distance, 324
Ohm, 250
Ohm's law, 271, 277, 283, 286
Optics, geometric, 323-349

Orbits, 178, 180, 184

Parallel circuits, 279-281
Parallelogram method, 82, 89-91, 240
Particles
  charged, in **B** fields, 220-223
  charged, in **E** fields, 207, 227, 264
Pendulum, 272-274
Period
  circular motion, 158, 160
  of wave, 300, 302
Phase, 310
  change of, 311
  points in, 310-312
  points out of, 310-312
  sources in, 310
Photoelectric effect, 305, 312-314
Photons, 304, 306, 314-316
Physics
  solving problems, 18-27
  symbols in, 16-18
  unique study habits, 1-3
  units in, 10-14
  use of formulas in, 7-10
  use of words in, 4-7
Potential difference, 249
  electric field intensity and, 253
  induced, 268
Potential energy, 249
  base level of, 252, 269
  elastic, 262-263
  electric, 253
  gravitational, 256-258, 273

nuclear, 297
Power, 249, 253-256, 260-262
Principal axis, 324-325
Projectiles, 159
  launched at an angle, 163-166, 169
  launched horizontally, 161-163
Proton
  in **B** field, 220-223
  in **E** field, 264

Quantum theory, 304

Radius of curvature, 326
Rarefactions, 307
Ray diagrams, 328,338-344
Real images, 328, 330, 338-340
Reference point, 57, 165
Reflection
  angle of, 324
  law of, 326-327
  total internal, 326,332
Refraction, 330-338
  angle of, 324
  index of, 324
  law of, 326-328
Resistance, 250
  effective, 269, 277, 279-282
Resolution
  of forces, 99-102
  of vectors, 81, 99-102
Resultant, 80
Rheostat, 276

Satellites, 180,184,190
Scalar quantities, 80
Scale, vectors to, 83
Series circuits, 276-279
Single slit, 305,316-318

Snell's law, 326-328, 330-338
Solenoid, 214
Solving problems,18-27
Sound, 307-309
Sources, in phase, 310-312
Spaceships, 243-244
Specific heat, 249
Springs, 262-263
Spring constant,249,262
Stopping potential, 301, 312-314
Superposition of waves, 300, 310-312

Temperature, 270
  heat and, 288-290
Tension, force of, 97, 125-128
Tesla, 216
Test charge, 195
Test magnet, 211, 224
Threshold frequency,301
Time, in formulas, 8,37
Total internal reflection, 326, 332
Two-dimensional motion, 157-177

Units, 10-16

Vectors, 80-114
  addition of,82,84-105, 200-202
  components of, 80
  resolution of, 81, 99-102
  subtraction of, 95-96
Velocity
  average, 38, 44-45
  escape, 178, 188-190
  instantaneous, 61
  negative, 58, 234

orbital, 178, 180, 184
uniform, 37,39,45,47
Virtual focal point, 325,329
Virtual image, 328, 330, 341, 344-345
Volt, 249
Voltage, induced, 268
Voltage drop, 269, 277

Watt, 249, 271
Wavelength, 299
  De Broglie, 307
Wave(s)
  boundary behavior of, 327,330
  characteristics of, 299-301, 303
  electromagnetic, 300, 305, 313
  matter, 305, 307
  superposition of, 300, 310-312
  the, nature of light, 304
Webers, 211, 226
Weight, 116, 118
Weightlessness, apparent, 181
Words, in physics, 4-7
Work, 248-251
  net, 251, 262
  energy and, 251-252
Work function, 301